Photonic Applications for Radio Systems and Networks

For a listing of recent titles in the
Artech House Applied Photonics Series,
turn to the back of this book.

Photonic Applications for Radio Systems and Networks

Fabio Cavaliere
Antonio D'Errico

ARTECH HOUSE
BOSTON | LONDON
artechhouse.com

Library of Congress Cataloging-in-Publication Data
A catalog record for this book is available from the U.S. Library of Congress.

British Library Cataloguing in Publication Data
A catalog record for this book is available from the British Library.

ISBN-13: 978-1-63081-665-0

Cover design by John Gomes

685 Canton Street
Norwood, MA 02062

10 9 8 7 6 5 4 3 2 1

Contents

CHAPTER 1

Introduction

For years, the evolution of optical networks and mobile networks followed independent paths. This is not surprising, considering the differences in both type of carried traffic and propagation channel physics. Mobile systems were originally designed to collect voice traffic, characterized by strict timing requirements and low capacity, over the small coverage area of a cell. Data traffic prevailed in mobile systems only after the relatively recent introduction of the fourth mobile system generation, 4G or Long-Term Evolution (LTE). Optical transmission systems are natively conceived instead as large pipes where heterogeneous traffic flows (voice traffic from the mobile network, Internet data from residential users of fixed access networks, business traffic from enterprises, etc.) convey and are transported over distances as far as hundreds or thousands of kilometers. Even the physics of the propagation channel could not be more different. Air is an isotropic, nondispersive propagation medium, subject to significant fluctuations caused by variable weather and multipath interference. Optical fiber is a guided medium—static in essence—where dispersive and nonlinear propagation effects are dominant, especially over long distance. All this led the transmission technologies for mobile and optical systems to diverge. Mobile systems rely on adaptive techniques based on sophisticated digital signal processing (DSP) algorithms, which until a few years ago could not be extended to fiber transmission; this was not only due to the different channel model, but because the high capacity in optical fiber would have required switching speeds not attainable with the available digital integrated circuits (ICs). Thus, the design of optical transmission systems was basically an analog design based on creative solutions that were developed ad hoc, without design standards like the one developed for ICs. As an important consequence of that, photonics could never reach the economies of scale and consumer pervasiveness that electronics have; it remained confined in physicists' labs or network operators' exchanges, with a few important exceptions such as the light-emitting diodes (LEDs) used in TVs or smartphones.

The explosion of the Internet dramatically changed the picture. Initially, high speed and reliable connectivity was all it was required to connect to the Internet, with content and data processed and managed locally within PCs connected to the wireline network. Gradually, content and data processing moved into the Cloud and datacenters, requiring unprecedented data storage capabilities and processing speeds, challenging the IC designers to further boost the Moore law. Such advances in electronics memories and processors had important consequences on both mobile and optical networks. In optical networks, they enabled the application of digital signal processing techniques to optical transmitters and receivers: Today, it is possible to send 100 Gbit/s on a single optical carrier, which was previously unconceivable. However, consequences were more disruptive on the mobile network, and are still shaping our everyday life. What was conceived as a portable phone is today a universal device, supporting not only voice calls, but the ability to view and edit documents, take and store photos, shoot videos — and exchange it all with the cloud and other devices/users. So, two peculiar features of the optical network — the support of different type of traffic and the high transmission capacity — became requirements of the fifth mobile network generation (5G). As a domino effect, the capacity required by the optical network increased as well, starting from the first access and aggregation stages. However, supporting an aggregate capacity of several hundred gigabits or a few terabits per second is not easy in access, which is a network segment much more sensitive to cost than the backbone, where such capacity values are instead common and require special technologies and design methods, which are the subject of this book.

Additional requirements inherited by the mobile network, such as the need to guarantee a low latency, make the challenge even more arduous for the optical network designer, asking them to acquire competence in mobile system engineering. But vice versa, those who plan a mobile network cannot ignore anymore the boundary constraints imposed by the fiber network that risk being a serious design bottleneck. In a few words, mobile and optical networks should be jointly designed. Unfortunately, the independent evolution of the optical and mobile networks led to a parallel separation of education and career paths, which only have a basic background in communications engineering in common. This is an important gap that this book intends to fill, at least partially. Both of the authors have a background in optical communications and photonic technologies, but also have the chance to work in one of the biggest companies active in the research and development of mobile communication systems. In addressing the issues of designing an optical transport network for 5G, they would personally experience the disorienting impact of interacting with colleagues speaking a different language and seeing things from a different perspective. Just to make a couple of examples, optical network designers tend to underestimate hardware costs and power consumption issues, which are crucial in a mobile network; however, wireless network designers are not familiar with design constraints, such as those imposed by the presence of deployed fiber infrastructure that must be reused, or the complex multilayer architecture of an optical network where wavelength routing, circuit multiplexing and packet switching coexist to provide the network with a flexibility that cannot be ensured by photonic technology alone. After the first impact, the authors experienced new exciting challenges and learning opportunities; this is firsthand experience they would like to transmit to the reader. Excellent specialized books already exist on both wireless and optical networks, but the intention of this work is to help people to gain basic complementary skills in both the areas.

This book intends, on one hand, to provide optical engineers working in operator or service provider companies, telco equipment industries, and research institutes with practical design guidelines for the next-generation mobile transport architecture. On the other hand, those who are experts in wireless communications are often not aware of the many possibilities offered by photonics, such as optical beamforming and optical processing of radio carriers — appealing techniques when moving to millimeter waves, where electronics start suffering from bandwidth limitations. In both cases, this book will provide a system-level description of the optical building blocks and will give the strictly necessary detail on the technology. The goal of the book is to provide engineers, professionals, and researchers with practical system design rules and guidelines for adopting a top down perspective, which starts from application scenarios, to the technologies enabling those scenarios. Unnecessary details about device physics or too much specialized technical content is avoided to make the book accessible to people with different technical background and with graduate skill level.

It has already been mentioned that the virtualization and centralization of mobile services is leading to the development of equipment (e.g., baseband processing units and switches) able to manage terabit/s capacities in a single board, which, however, shall have to introduce a limited increase of size and energy consumption compared to the current technology. New, highly dense hardware platforms based on integrated photonics that are helping to improve loss and power consumption of the current baseband units and packet switches will also be addressed in this book. A third class of target readers consists of a new generation of hardware engineers, skilled in both electronics and optics, capable of designing Opto-ICs, where electronics and photonics coexist and are wisely mixed to achieve the best speed and energy efficiency performance.

The book is divided into 10 chapters. Chapter 2 describes the air interface of 4G and 5G systems, introducing concepts such as orthogonal frequency division multiplexing (OFDM) access, time-frequency resource grids, interference mitigation techniques, time-duplex and frequency-duplex frame structures, multiantenna systems, and signal processing chains. The radio access network architecture and its requirements are discussed in Chapter 3, which starts with 5G use cases and requirements, and then describes the radio protocol stack, with special emphasis on the HARQ protocol and its consequences on the latency budget of a mobile communication system. The split radio access network architecture is discussed, as well as the main functional split options, and logical interfaces and deployment scenarios for fronthaul and backhaul applications. Bit rate and latency are calculated for the split options of common use. Chapter 4 illustrates the transmission modeling techniques in a radio access optical network. The main linear and nonlinear propagation effects in optical fiber are presented, with their impact on the system performance. All the propagation penalties can be calculated using a spreadsheet without requiring complicated numerical modeling, in order to provide the designer with simple rules of thumbs for network planning.

Chapter 5 illustrates optical technologies, such as modulation formats and optical amplifiers, used in digital radio access networks based on different deployment scenarios: point-to-point fiber systems, wavelength division multiplexing systems, and passive optical networks. Due to their importance and complexity, optical switching technologies are discussed in Chapter 6, addressing various implementations: Single core and multicore fiber, multiwavelength optical switches, and others. New integrated photonic technologies for optical switching are also discussed.

Chapter 7 discusses how optical technologies can directly impact the design of mobile network and equipment, independently of the presence of a fiber transport network. It describes technologies, such as directly modulated lasers and external optical modulators that are applicable to analog fronthaul systems, also known as radio over fiber (RoF), and provides the design rules for such systems, for both single or multiple subcarriers. Chapter 8 illustrates the principles of optical beamforming and discusses issues in providing accurate phase and frequency control of the antenna elements. Finally, Chapter 9 introduces integrated optical interconnection technologies for new-generation radio systems, such as optointegrated circuits and optical printed circuits boards. Chapter 10 provides a summary of each previous chapter.

Before concluding, the authors would like to thank Claudia and Maria for their constant support. Moreover, they are grateful to all Ericsson colleagues that made this work possible, inspiring them with continuous stimulating inputs. Listing them individually would require a separate book. Special thanks to: Pierpaolo Ghiggino, to who one of the authors owes the technical and business skills that made this book possible; Ernesto Ciaramella, who inspired one of the authors with unwavering support in his research studies; Antonella Bogoni, Luca Potì, and Paolo Ghelfi , who shared with the authors a long research path; Enrico Forestieri, who not only provided some of the results shown in this book but also, and especially, taught to the authors to look beyond surface appearances; all the researchers who worked with the authors on pioneering research projects in optics and photonics; and many others whose help was invaluable.

CHAPTER 2

Radio Systems Physical Layer

2.1 Introduction

The advent of mobile radio communication systems, with the concurrent explosion of the Internet, was definitely one of the major technological revolutions of the human history. Mobile devices connected to the Internet help, and sometimes steer, our lives to such an extent that nobody could imagine their life today without being connected to a smartphone or laptop. According to [1], the number of mobile subscribers in 2018 was around 5.4 billion, with a growth rate of about 52% year-on-year, though with large differences between countries.

From the early 1990s, there were five successive generations of radio systems. The first generation (1G) was based on analog standards like the Total Access Communication System (TACS) in Europe and the Advanced Mobile Phone System in North America. The second generation (2G) was based on the Global System for Mobile Communications (GSM) digital standard, and its evolutions, the General Packet Radio Service (GPRS) and Enhanced Data rates for GSM Evolution (EDGE) for data transmission over packets. The third generation (3G) used the Universal Mobile Telephone System (UMTS) standard, based on code division multiplexing access (CDMA), which evolved into the HSDPA—High Speed Downlink Packet Access—standard. The recent 4G-LTE was designed to support high speed data transfer and will be extensively illustrated in this chapter, as well as the next-generation 5G, designed to fit the requirements of broadband access networks supporting a diverse range of new applications like robotic surgery, virtual reality classrooms, and self-driving cars.

There are three broad families of use cases for which 5G wireless access has been developed: Enhanced mobile broadband (eMBB), massive machine-type communications (mMTC) and ultra-reliable low-latency communications (URLLC). eMBB provides mobile broadband access at an extremely high data rate, low latency, high user density, and wide coverage. mMTC is designed for massive Internet of Things (IoT) scenarios, where hundreds of thousands of low cost and battery powered devices per km^2 are connected. Typical mMTC applications are smart metering, logistics, and body sensors. Finally, URLLC will make communicating between devices possible with high reliability and very low latency, which is essential in applications such as vehicular communication, industrial control, factory automation, remote surgery, smart grids, and public safety. To meet the requirements of these three diverse scenarios, a new standard [2] known as New Radio (NR) was developed by the 3rd Generation Partnership Project (3GPP).

In 5G, the radio carrier frequency spans over a wide range, from approximately 1 GHz to 100GHz, with lower carrier frequencies primarily used in macrosites to provide more coverage and higher carrier frequencies that are mainly used in microsites and picosites having a smaller coverage area. As in LTE, licensed spectrum will be exploited to provide high service quality and reliability, while unlicensed spectrum will be primarily used to provide additional capacity.

The development of the NR standard started at 3GPP in 2016, with the purpose of making it commercially available before 2020. A first standardization phase, specifying its essential features, was completed in 2018, followed by a second phase, addressing the full set of requirements defined by ITU-R for the next generation of mobile communication systems, called International Mobile Communications 2020 (IMT-2020). The backward compatibility of the NR standard with LTE is not strictly required, however, it was designed to ensure to the mobile network operates a smooth evolution from their current LTE deployments.

Section 2.2 will provide an overview of the physical layer specifications of a 5G mobile system, focusing on the aspects that affect the requirements of the underlying optical transport network. However, due to the similarities with LTE in both used nomenclature and technology, it is opportune to describe the 4G standard, first.

2.2 Physical Layer of 4G Radio Systems

In the LTE standard, the user device is called UE, and the base station is called eNodeB. The transmission directions from eNodeB to UE and from UE to eNodeB are referred as downlink (DL) and uplink (UL), respectively. The radio interface between UE and eNodeB specifies the transfer of both signaling and user data. It uses orthogonal frequency division multiplexing (OFDM and orthogonal frequency division multiple access (OFDMA) as transmission and multiple access technologies in the downlink. In the uplink, a slightly different approach, named single carrier frequency division multiplexing access (SC-FDMA), is adopted. Those techniques make a flexible allocation of bandwidth (BW) and frequencies possible, allowing the operators to configure the radio link according to the available frequency spectrum.

2.2.1 Orthogonal Frequency Division Multiplexing

Subcarrier multiplexing (SCM) and OFDM are two examples of frequency division multiplexing (FDM). An SCM signal modulating a radio carrier consists of the sum of individually modulated tones (subcarriers), narrowly spaced in frequency and with independent phases. To avoid interference, the BWs of the subcarriers do not overlap. Due to this, and because of the finite accuracy the subcarrier frequencies are generated with, frequency gaps [guard bands (GBs)] are usually present between adjacent subcarriers. Figure 2.1 shows an example of an SCM spectrum with five subcarriers, centered at the frequencies f_1, ..., f_5. The BW is equal for all the subcarriers, and adjacent subcarriers are separated by a GB.

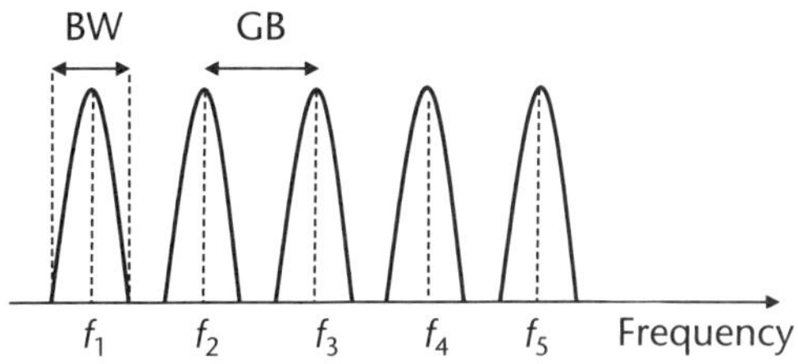

Figure 2.1 Subcarrier multiplexing spectrum.

Signals that do not overlap in frequency, like the subcarriers of an SCM signal, are an example of orthogonal signals. Another example is given by signals that are transmitted over separated time intervals and do not overlap in time. In general, two complex signals $x_1(t)$ and $x_2(t)$, defined over a time interval (a, b), are said to be orthogonal when their inner product [see (2.1)] is zero [3].

$$\langle x_1(t), x_2(t)\rangle = \int_a^b x_1(t) \cdot x_2^*(t)\,dt \tag{2.1}$$

A digital communication system is said to adopt an orthogonal modulation format when the bits-to-transmit are partitioned in blocks, and the block are mapped into orthogonal waveform. OFDM is an example of an orthogonal modulation format. It can be considered as a spectrally efficient variant of SCM, where the subcarriers are allowed to overlap, and the GBs are not necessary. An example of OFDM spectrum is shown in Figure 2.2. Comparing it with the SCM spectrum of Figure 2.1, the gain in spectral efficiency can be appreciated. The subcarriers are spaced by $1/T$. For reasons that will be clearer at the end of this paragraph, T is called symbol time.

Using the lowpass equivalent, or baseband, signals representation [4], an OFDM signal with N subcarriers, defined over the time interval (0, T), can be written as

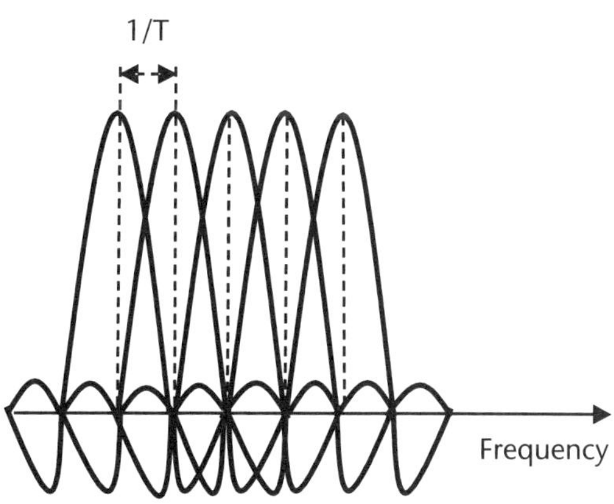

Figure 2.2 Principle representation of the OFDM spectrum.

$$s(t) = \sum_{k=0}^{N-1} S_k e^{j2\pi \frac{kt}{T}} \quad (2.2)$$

where T is the symbol time and S_k is the modulation symbol associated to the kth subcarrier, $f_k = \frac{k}{T}$. Equation (2.1) can be used to verify that $e^{j2\pi \frac{kt}{T}}$ and $e^{j2\pi \frac{ht}{T}}$ are orthogonal over (0, T) when $h \neq k$. The time samples, s_m of $s(t)$ at the instants $t_m = \frac{mT}{N}$ can be calculated as the inverse fast Fourier transform (IFFT) of the modulation symbols S_k:

$$s_m = \sum_{k=0}^{N-1} S_k e^{j2\pi \frac{km}{N}} \quad (2.3)$$

Conversely, the symbol S_k can be derived from the time samples s_m by means of a fast Fourier transform (FFT):

$$S_k = \frac{1}{N} \sum_{m=0}^{N-1} s_m e^{-j2\pi \frac{mk}{N}} \quad (2.4)$$

Equation (2.4) is also known as the frequency domain representation of an OFDM signal. Equations (2.3) and (2.4) suggest a procedure to generate an OFDM signal by means of DSP techniques. The first step of the procedure is to encode information bits into modulation symbols using, for example, quadrature amplitude modulation (QAM) [5]. The bit rate can be individually adjusted on each subcarrier by using QAM constellations of different size. For example, a 16-QAM subcarrier could switch to 4-QAM, halving the bit rate, if affected by selective fading. In a second step of the procedure, time domain samples are generated by calculating the IFFT of the modulation symbols [see (2.3)]. The time domain samples are sent to a digital-to-analog converter (DAC) with two outputs, corresponding to real part and imaginary part of the OFDM signal in (2.2). The two signals at the output of the DAC are sent to the inputs of an IQ modulator, named for its in-phase (I) and in-quadrature (Q) ports, to perform up-conversion at the carrier frequency. Additional pulse shaping and filtering stages may be performed, using either digital or analog techniques.

Due to reflections, scattering, or atmospheric effects, several delayed and attenuated copies of a transmitted radio signal can arrive to the same receiver. The effect is known as multipath propagation and is a typical source of interference in radio transmissions [6]. In general, data and interfering signals will have different amplitudes, but this is an aspect not essential to the following discussion. Figure 2.3(a) shows received data and interfering OFDM signals, where adjacent symbols partially overlap due to multipath interference. For example, the final part of Symbol 1 in the data signal, within the time interval between t_2 and $t_2 + \Delta T$, cannot be correctly received because it overlaps with the initial part of Symbol 2 of the interfering signal. This effect is known as intersymbol interference (ISI).

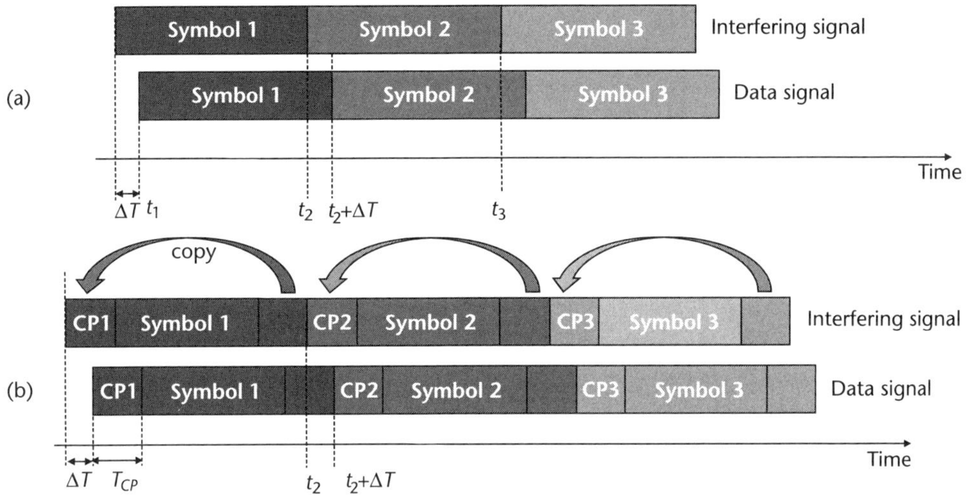

Figure 2.3 Signal experiencing (a) multipath interference and (b) interference mitigation by means of cyclic prefix.

Within the time intervals where time-shifted copies of the same symbol are received, as happens for Symbol 2 between $t_2 + \Delta T$ and t_3, digital equalization techniques [7] can be used to recover the signal. In principle, the ISI could also be compensated by means of such techniques, such as using a transversal finite impulse response (FIR) equalizer [8], for example. However, in the presence of ISI spanning over multiple symbol times, the number of delay taps and multipliers of the equalizer would excessively increase, along with introduced delay and complexity. For this reason, in an OFDM transmission system, the ISI is usually compensated by appending a cyclic prefix (CP) to the symbols. In the following discussion, ΔT will indicate the upper bound of the relative delay between data signal and interfering signal, which is supposed to be known. The insertion of the CP consists of copying the last part of each symbol at the beginning of the symbol itself, as shown in Figure 2.3(b). The duration of the copied part must equal or higher than ΔT: This is the time length of the CP, T_{CP}, a parameter which is usually determined by system design. After the introduction of the CP, even if the final part of Symbol 1 experiences interference from Symbol 2, it can still be retrieved at the beginning of Symbol 1, where it has been copied. The CP has the disadvantage of adding an overhead to the signal, but it is widely adopted anyway due to its implementation simplicity. Because of the periodic nature of the FFT, a delay in the time domain corresponds to a rotation in the frequency domain, which introduces a complex multiplicative term $e^{j2\pi\frac{m_{cp}k}{N}}$ in (2.4), with $T_{CP} = \frac{m_{CP}T}{N}$. At the receiver, after having compensated the rotation by multiplying the signal by $e^{-j2\pi\frac{m_{cp}k}{N}}$, the CP can be simply discharged.

Besides ISI, FDM systems can suffer from intercarrier interference (ICI). ICI occurs when the BWs of adjacent subcarriers overlap. Nominally, an OFDM signal is immune from ICI, because at the peak frequency of each subcarrier, the spectra of the other subcarriers are equal to zero (Figure 2.2). Indeed, an OFDM signal can be considered as an SCM signal with $1/T$-spaced subcarriers [see (2.2)]. Since the symbol time is T, the pulses on each subcarrier are shaped with ideal rectangular functions, equal to 1 within the symbol time and 0 elsewhere. Hence, due to the

Fourier transform properties, the spectrum of the kth subcarrier is shaped as sinc(f – k/T), with $sinc(x) = \frac{sin(\pi x)}{\pi x}$. The function $sinc$(x) is null for any integer argument, except $x = 0$, at which it equals 1. Consequently, the OFDM spectrum at the frequency k/T has no contribution, except that of the kth subcarrier, as it is visualized in Figure 2.2. An exact mathematical derivation of the OFDM power spectral density, including the CP, can be found in [9]. In real systems, several factors can break the orthogonality between the subcarriers and lead to ICI: Rapid time variation of the propagation channel, Doppler shift (especially in high mobility applications), phase noise, among others. Various methods exist for interference mitigation, but their description is out the scope of this book: A survey can be found in [10].

2.2.2 Orthogonal Frequency Division Multiplexing Access

The greatest advantage of using OFDM in a mobile system is the possibility to assign to each user dedicated time and frequency resources. This is illustrated in Figure 2.4, where frequencies (i.e., groups of OFDM subcarriers) and time slots are assigned to different users, indicated by different colors. This is the OFDMA approach.

The time and frequency resources are arranged in units, called scheduling blocks or physical resource blocks (PRBs), which will be defined quantitatively in the next paragraphs. They correspond to the rectangular elements of the time-frequency diagram in Figure 2.4, called a resource grid. At the eNodeB, scheduling mechanisms assign PRBs to the users in such a way that it is possible to dynamically schedule the users on the frequencies they experience the best quality on. The reference time interval in LTE radio transmissions is called transmission time interval (TTI) and equals 1 millisecond. Each row of the resource grid in Figure 2.4 corresponds to one TTI. Different sets of subcarriers are assigned to different users at time intervals in multiples of the TTI.

The approach described so far is used in the downlink. In LTE (but not in 5G), the uplink uses a slightly different technique, called SC-FDMA or discrete Fourier transform spread OFDM (DFTS-OFDM). As for OFDMA, in SC-FDMA, data bits are first converted into subcarrier constellation symbols using modulation formats like binary phase shift keying (BPSK), quadrature phase shift keying (QPSK) or 16-QAM. Instead of using those symbols in the IFFT for generating an OFDM signal, according to (2.3), an intermediate step is performed. In this step, blocks of consecutive symbols are converted by using a discrete Fourier transform (DFT) into contiguous discrete subcarriers—which are a subset of the OFDM subcarriers generated by the IFFT in (2.3)—in such a way that the modulation symbols on those contiguous subcarriers are no longer independent, leading to a reduction of

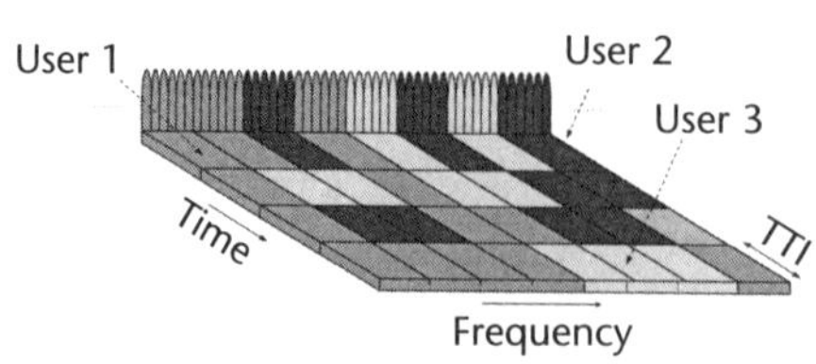

Figure 2.4 OFDMA resource grid.

the peak-to-average power ratio (PAPR), which is an issue in OFDM transmissions [11]. A lower PAPR in uplink makes the UE more power efficient and cost effective. SC-FDMA will not be discussed further since it was abandoned in 5G, which uses CP OFDM in both downlink and uplink.

2.2.3 LTE Frame Structure

LTE envisages both frequency division duplexing (FDD) and time division duplexing (TDD). In FDD systems, downlink and uplink operate at different carrier frequencies. In TDD systems, downlink and uplink use the same frequency but transmit at different time intervals, separated by a GB. FDD is efficient in the case of symmetric traffic, because it does not waste bandwidth during the switchover between transmitter and receiver. Moreover, as base stations transmit and receive in different subbands, they normally do not interfere with each other. TDD has a strong advantage when the the uplink and downlink data rates are asymmetric, as the communication capacity can be dynamically allocated to uplink and downlink, proportionally to the actual amount of data. The frame structure of an FDD system is simpler than that of a TDD system and will be used in the following to illustrate how the frame is assembled in an LTE system, starting from the modulation symbols. The time domain structure of the FDD frame is shown in Figure 2.5 and is valid for both downlink and uplink. One 10-ms frame consists of ten 1-ms subframes. Each subframe consists of two 0.5-ms slots and each slot consists of 7 OFDM symbols, with their own CP. Another frame structure with only 6 OFDM symbols with a longer CP is defined by the LTE standard, but it will not be discussed here.

A PRB in the OFDMA resource grid, illustrated in Section 2.2.2, is defined as consisting of 12 consecutive subcarriers transmitted within one 0.5-ms slot (Figure 2.6). A resource element (RE) is defined instead as one subcarrier over one OFDM symbol. The frequency spacing, Δf, between OFDM subcarriers is 15 kHz, so that the BW of one PRB is $12 \cdot \Delta f = 180$ kHz.

According to (2.2), the symbol time T equals the inverse of the subcarrier spacing $T = 1/\Delta f$ and is approximately 66,67 μs. According to the Fourier transforms theory, a time domain signal, defined over the interval $(0, T)$ and sampled at the

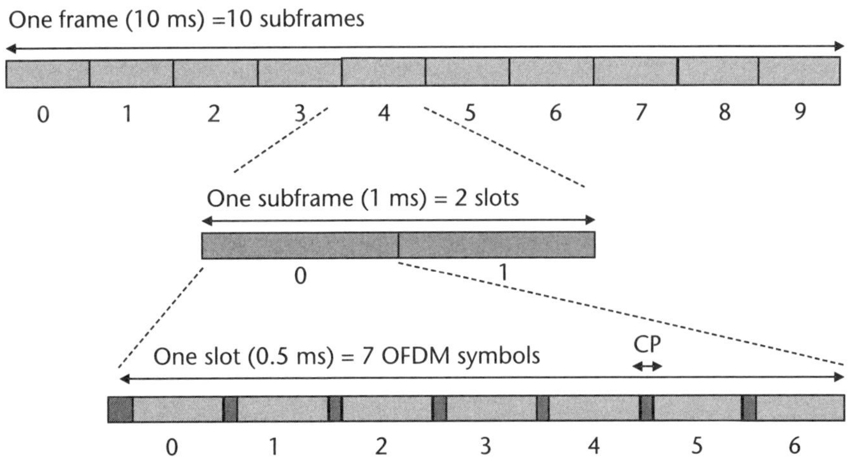

Figure 2.5 FDD frame structure in LTE.

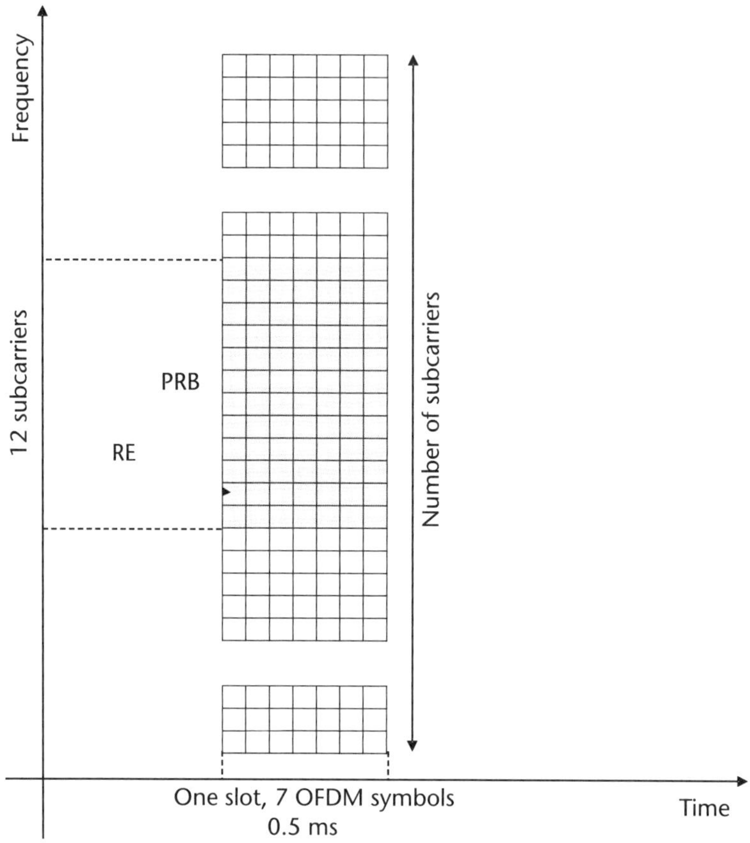

Figure 2.6 LTE resource grid.

time instants $t_m = m\frac{T}{N}$, corresponds to a frequency domain signal, defined over the frequency range (0, N/T) and sampled at the discrete frequencies $f_k = \frac{k}{T}$, with both m and k taking integer values from 0 to N–1. For the OFDM signal in (2.2), time and frequency domain samples are given by (2.3) and (2.4), respectively. The interval, T/N, between two consecutive time domain samples is called sampling time and its inverse, f_S, is called sampling frequency. Due to the Fourier transform properties mentioned above, a simple relation exists between sampling frequency f_s, number of subcarriers (or FFT size) N, and subcarrier spacing Δf:

$$f_s = \left(\frac{T}{N}\right)^{-1} = N \cdot \Delta f \tag{2.5}$$

In the LTE standard, the FFT size can take the values of 128, 256, 512, 1,024, 1,536, and 2,048. Using (2.5), the sampling frequency with N=2,048 is 30.72 MHz. The number of samples on each slot can be easily calculated as:

$$Samples_{slot} = f_s \cdot T_{slot} \tag{2.6}$$

Since T_{slot}=0.5 ms, with N=2,048, $Samples_{slot}$=15,360. The samples for each OFDM symbol are:

$$Samples_{symbol} = f_s \cdot T = N \tag{2.7}$$

The difference between the total number of samples per slot and the number of samples taken by the OFDM symbols is used for the CPs:

$$Samples_{CPs} = Samples_{slot} - (Symbols\_per\_slot) \cdot Samples_{symbol} \tag{2.8}$$

For N=2,048 and 7 symbols per slot, $Sample_{CPs}$=1,024: Since 1,024 is not a multiple of 7, the CP of the first OFDM symbol (symbol 0 in Figure 2.5) is slightly longer (160 samples) than the CPs of slots 1-6, which last 144 samples. The number of CP samples at different FFT sizes can be calculated, observing that the right side of (2.8) is proportional to N, as can be verified by expressing it as function of the sampling frequency given by (2.5):

$$Samples_{CPs}(N) = N \cdot (\Delta f \cdot T_{slot} - Symbols\ per\ slot) \tag{2.9}$$

The time allocated to all CPs in the frame, T_{CPs}, is obtained multiplying (2.9) by the sampling time f_s^{-1}. As demonstrated by (2.10), it is independent of the FFT size and equal to 33.33 μs.

$$Time_{CPs} = T_{slot} - \frac{Symbols\ per\ slot}{\Delta f} \tag{2.10}$$

Hence, proportionally scaling the results obtained for N=2,048, the duration of the CP is (160/1,024)·33.33 μs = 5.21 μs for the first symbol in the slot and (144/1024)·33.33 μs = 4.69 μs for the remaining six symbols. The number of samples required by each CP for different values of N can be calculated by multiplying these values for the sampling frequency, using (2.5).

2.2.4 LTE Systems Bandwidth

Six different values of BW are defined by the LTE standard: 1.25, 2.5, 5.0, 10.0, 15.0, and 20.0 MHz. The FFT size, N, is proportional to the BW, with N=2,048 for BW=20 MHz [see (2.11)]:

$$N = 2048 \frac{BW(MHz)}{20} \tag{2.11}$$

Not all N subcarriers are used for data transmission: Some of them are set to zero to mitigate the ICI, a technique known as zero padding. For example, with N=1,024 and BW=10 MHz, 423 subcarriers are zero padded and the remaining 601 are used for data. The usual practice is to zero pad some central subcarriers: Due to the periodic nature of the FFT, this is the equivalent to generating a lowpass modulating signal. An alternative zero padding rule is to only fill the central subcarriers. The subcarriers used for data are grouped in PRBs, as explained in Section

2.2.3. The number of PRBs, N_{PRB}, is proportional to the BW, with N_{PRB} set to 100 at a BW of 20 MHz, as in (2.12):

$$N_{PRB} = 100\frac{BW\,(MHz)}{20} \tag{2.12}$$

The number of occupied, nonzero padded subcarriers, N_{actual}, is calculated in (2.13), where the number of subcarriers per PRB is 12 (Figure 2.6) and the addend 1 at the right side takes into account of the direct current (DC) component, which is not considered to be part of the PRBs.

$$N_{actual} = Subcarriers\ per\ PRB \cdot N_{PRB} + 1 \tag{2.13}$$

For example, with BW=20 MHz, N_{PRB}=100 and N_{actual}=1,201. The actual value of bandwidth, BW_{actual}, is calculated in (2.14), multiplying the number of occupied subcarriers by the subcarrier frequency spacing, Δf:

$$BW_{actual} = N_{actual} \cdot \Delta f \tag{2.14}$$

Equation 2.14 gives the actual bandwidth on air of an LTE signal, which is lower than the nominal BW. For example, a nominal BW of 20 MHz, which is the value commonly referred to in the technical literature as the bandwidth of an LTE system, corresponds to an actual BW of 18.015 MHz. Finally, the number of zero-padded subcarriers can be calculated as:

$$N_{zero\ pad} = N - N_{actual} \tag{2.15}$$

It is 847 at a BW of 20 MHz (i.e., N=2,048).

2.2.5 TDD Frame Structure

The frame structure of a TDD system is similar to the FDD frame shown in Figure 2.5, with the following modifications introduced to allow the switchover between DL and UL without loss of data. The TDD 10-ms frame is shown in Figure 2.7: It is divided in two 5-ms half-frames, each half-frame consisting of five 1-ms subframes. In Figure 2.7, time intervals used in DL and UL are indicated by the letters D and U, respectively.

Within each half frame, the second subframe (subframes 1 and 6 in Figure 2.7) has a special structure: It consists of a DL part (downlink pilot time slot, DwPTS), a guard period (GP), and an UL part (uplink pilot time slot, UpPTS). The downlink-to-uplink switching point always takes place within the second subframe of each half-frame, and there can be two of such switching points within the frame. The corresponding uplink-to-downlink switching point can take place at any subsequent subframe boundary within the half frame. Note that due to this mechanism, the first subframe of each half-frame is always used for DL transmission. Several

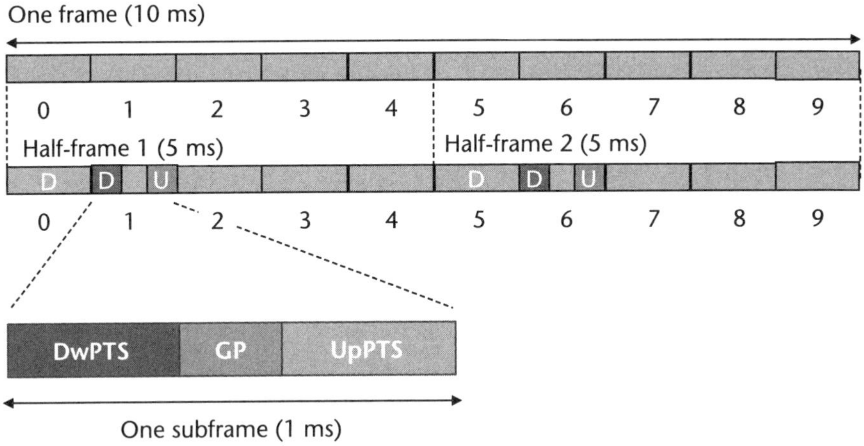

Figure 2.7 TDD frame structure.

frame configurations are possible, allowing to flexibly allocate UL and DL capacity according to the traffic demand. Examples of configuration are reported in Figure 2.8.

In the special subframes, the relative length of DwPTs, GP and UpTS fields can be configured, as shown in Figure 2.9. The numbers within each field indicate the number of symbols in the different configurations (note they always add up to 14, as expected).

2.2.6 LTE Physical Layer Parameters

The physical layer parameters of an LTE system, which are independent of the BW value, are summarized in Table 2.1.

Table 2.2 lists instead the most significant bandwidth dependent parameters, calculated as illustrated in the previous paragraphs.

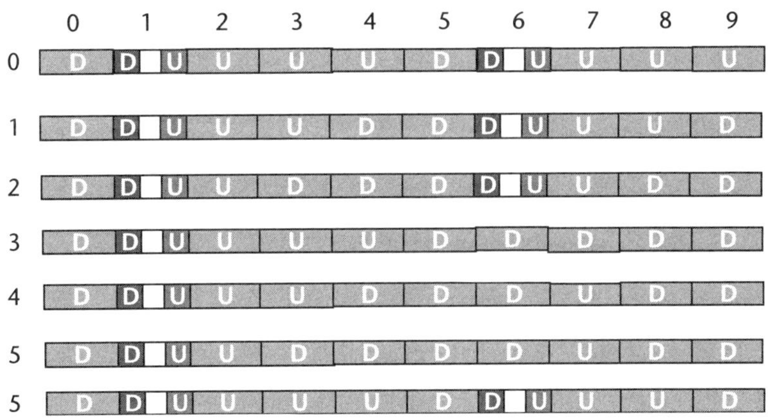

Figure 2.8 TDD frame configuration.

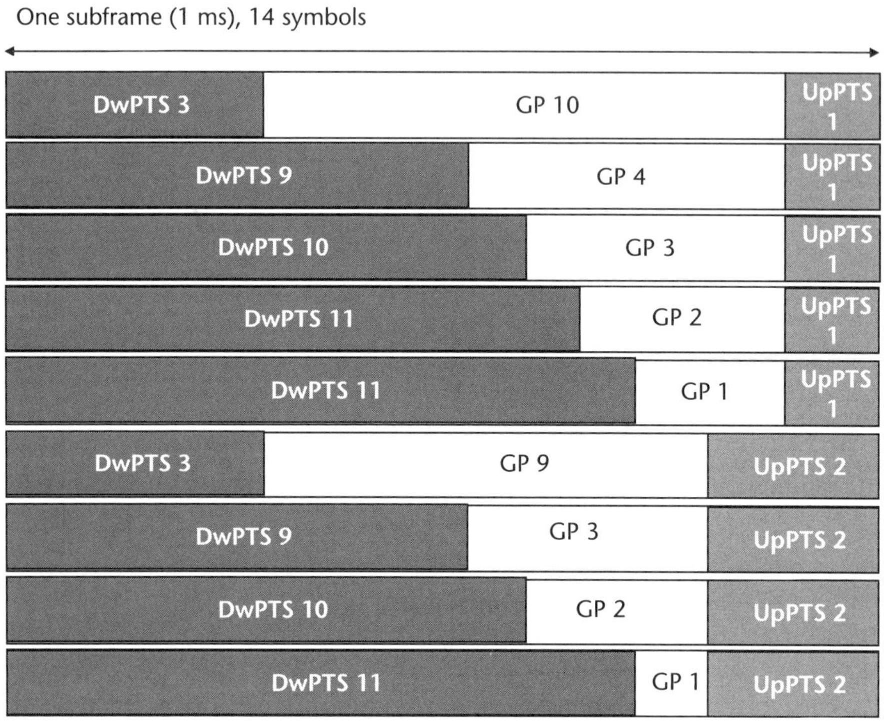

Figure 2.9 Configuration options of TDD special subframes.

Table 2.1 Bandwidth-Independent LTE Physical Layer Parameters

Frame Duration	10 ms
Subframe Duration	1 ms
Slot Duration	0.5 ms
Number of Symbols Per Slot	7 (normal CP), 6 (long CP)
Subcarrier Spacing	15 KHz
Symbol Time, Excluding the CP	66.67 μs
CP Duration	Normal CP: 5.21 μs (Symbol 0), 4.69 μs (Symbols 1–6)
	Long CP: 16.67
Subcarriers Per PRB	12
PRB Bandwidth	180 kHz

2.3 Physical Layer of 5G Radio Systems

The 5G standard, hereinafter simply referred as NR, resumes many aspects of the previous 4G generation. As in LTE, blocks of data bits are mapped into symbols according to some single carrier modulation format like QAM, and the symbols generate OFDM subcarriers by means of an IFTT, as shown in (2.3). The concepts of physical resource block and OFDMA resource grid are also reused. Since the LTE standard is supposed to be known from the previous paragraphs, the remainder

Table 2.2 Bandwidth-Dependent LTE Physical Layer Parameters

Nominal BW	*1.25 MHz*	*2.5 MHz*	*5 MHz*	*10 MHz*	*15 MHz*	*20 MHz*
FFT Size	128	256	512	1024	1536	2048
Sampling Frequency	1.92 MHz	3.84 MHz	7.68 MHz	15.36 MHz	23.04 MHz	30.72 MHz
Samples per Slot	960	1920	3840	7680	11520	15360
Samples per Normal CP	10 (#0) 9 (#1-6)	20 (#0) 18 (#1-6)	40 (#0) 36 (#1-6)	80 (#0) 72 (#1-6)	120 (#0) 108 (#1-6)	160 (#0) 144 (#1-6)
Samples per Long CP	32	64	128	256	384	512
Number of PRBs	6	12	25	50	75	100
Occupied Subcarriers	76	151	301	601	901	1201
Zero Padded Carriers	52	105	211	423	635	847
Actual Bandwidth	1140 KHz	2265 kHz	4515 kHz	9015 kHz	13515 kHz	18015 kHz

of this chapter will primarily focus on the extensions and differences that NR has introduced with respect to LTE.

2.3.1 Modulation Schemes

Both LTE and NR support the modulation formats QPSK, 16-QAM, 64-QAM and 256-QAM. Higher-order QAM modulation, like 1,024-QAM, is also envisaged, though not frequently used. In NR, $\pi/2$-BPSK has been added to the list of modulation formats to reduce the PAPR in the uplink and improve the efficiency of the UE power amplifier. This is a useful feature, especially in mMTC applications that use battery powered devices and usually transmit at low data rates that do not require high spectral efficiency modulation. A regular BPSK modulation scheme maps one transmission bit in one symbol, applying a 0° or 180° phase shift to a sinusoidal carrier if the bit to transmit is 0 or 1, respectively. A BPSK modulated signal is simple to generate and detect, but it has two main drawbacks: Phase ambiguity and sharp phase transitions. Phase ambiguity is a consequence of the fact that the absolute phase of a carrier cannot be measured but only its difference with another wave, used as reference. Since the initial phase of the carrier is not defined and the channel introduces a random, slowly varying phase offset, it is impossible, without additional coding, to understand if the detected symbol corresponds to a 0° or 180° phase shift. In theory, the amplitude of a BPSK modulated signal is constant. In practice, because of the modulator nonideality, sharp phase variations, like those occurring at the transitions between 0° and 180°, lead to an undesired amplitude modulation, which requires power amplifiers with a higher dynamic range. Moreover, sharp variations of phase and amplitude enhance the side lobes of the spectrum, causing ICI. $\pi/2$-BPSK has no phase ambiguity and reduces the phase excursions [12]. To do so, it increments the carrier by 90° phases at every transmitted symbol, so that the information is carried by the phase transitions rather than the phase itself. Moreover, the phase variations are less pronounced, as well as the associated amplitude modulation.

2.3.2 5G Numerology and Frame Structure

NR uses CP-OFDM modulation in both downlink and uplink, although DFTS-OFDM (see Section 2.2.2) is still envisaged in the UL for coverage-limited scenarios without multiple input, multiple output (MIMO) (see Section 2.4). Having the same signal format in both directions simplifies the design in applications that are expected to become widespread in 5G and requires similar equipment at transmitter and receiver: Wireless backhauling and device-to-device communication are two examples. The main novelty of NR compared to LTE is the introduction of multiple values of subcarrier spacing. In LTE, the subcarrier spacing Δf is always equal to 15 kHz, while in NR it is parameterized with an integer number μ, called numerology, which determines not only the frequency spacing but also other physical layer specifications, such as bandwidth of PRB.

$$\Delta f = 2^{\mu} \cdot 15\ KHz \quad \mu = 0, 1, ..4 \tag{2.16}$$

It can be seen from (2.16) that Δf ranges from 15 to 240 KHz. Sampling frequency, f_s, and sampling time, $1/f_s$, depends on Δf as in LTE [see (2.5)]:

$$f_s = N \cdot \Delta f = N \cdot 2^{\mu} \cdot 15\ kHz \tag{2.17}$$

with μ=0 and N=2,048, the sampling frequency and sampling time are 30.72 MHz and 32.552 ns, respectively, as in LTE. In NR, as in LTE, a PRB is defined as 12 contiguous subcarriers transmitted in one slot, so that the PRB bandwidth, 12·f, ranges from 180 kHz (the LTE value, obtained with μ=0) to 2,880 kHz (μ=4). While in LTE there are always two slots per subframe (Figure 2.5), in NR, the number of slots within a 1-ms subframe depends on the numerology:

$$Slots\ per\ subframe = 2^{\mu} \tag{2.18}$$

Note that with μ=0, subframe and slot coincide, differently from LTE. The slot duration can be calculated from (2.18), as

$$T_{slot} = \frac{1\ ms}{Slots\ per\ subframe} = \frac{1\ ms}{2^{\mu}} \tag{2.19}$$

In NR, the number of OFDM symbols per slot is always 14, independently of the numerology (for μ from 0 to 2, NR has also an option with 7 symbols per slot and two slots per subframe, as in LTE, but it will not be further discussed). Therefore, since both slot duration and number of symbols per slot are twice the value they have in LTE, the duration of an OFDM symbol with μ=0 remains constant, that is, 66.67 μs. Of course, the symbol duration also scales with the numerology:

$$T = \frac{66.67\ \mu s}{2^{\mu}} \tag{2.20}$$

The choice of the numerology depends on various design aspects, including carrier frequency, required quality of service in terms of latency and throughput, and hardware impairments such as oscillator phase noise [13]. For example, wider subcarrier spacings and shorter symbols are more suitable for latency critical services (URLLC) and higher carrier frequencies in small coverage areas. Narrower subcarrier spacing can be utilized instead for lower carrier frequencies in large coverage areas and narrowband devices. A scaling factor equal to a power of 2 ensures that slots and symbols of frames assembled with different numerologies can be easily aligned in time, which is especially important in TDD systems. However, this feature must be guaranteed even after the introduction of the CP: This is not simple, because in NR as in LTE, one CP is longer than the other ones to fit the subframe length (see Section 2.2.3). The length of the CP in NR is calculated according to the following two rules, which allow alignment of two subframes with different numerology:

- The duration of every symbol at μ=0, including the CP, equals the sum of the durations of 2^{μ} symbols at a different numerology;
- Other than the first symbol in every 0.5 ms, all other symbols in 0.5 ms have the same length.

The resulting frame formats for different values of μ are illustrated in Figure 2.10, referring to an FFT size of 2,048. Note that the first CP is longer than the following normal CPs.

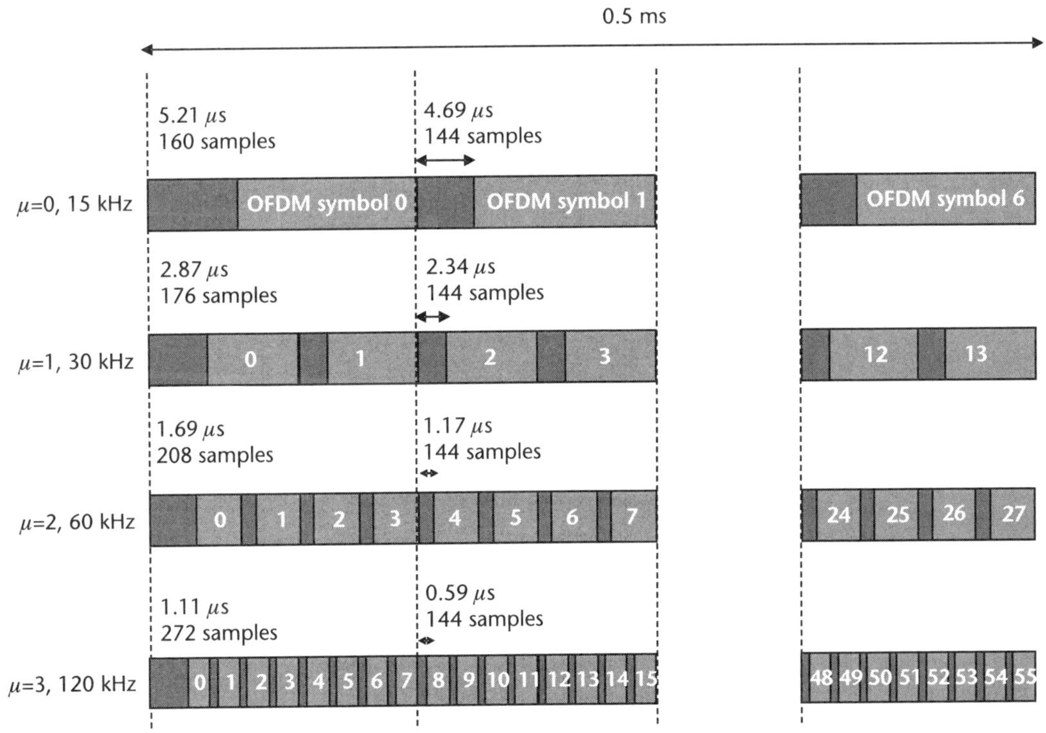

Figure 2.10 NR frame format at different numerologies.

The duration of a normal CP can be calculated scaling down the value at 15 kHz subcarrier spacing, which is 4.69 μs (see Section 2.2.3):

$$T_{CP} = \frac{4.69\ \mu s}{2^{\mu}} \tag{2.21}$$

The number of samples of the CP, N_{CP}, is the product of CP duration [see (2.20)] and sampling frequency [see (2.17)]:

$$N_{CP} = f_s \cdot T_{CP} = 4.69\ \mu s \cdot 15\ kHz \cdot N = 0.07035 \cdot N \tag{2.22}$$

Note that the CP length is independent of the numerology. For N=2,048, N_{CP} equals 144, as for LTE. The number of samples of the first CP, N_{CP0}, can be calculated imposing the time alignment of the end of the first symbol of the μ-1 numerology frame with the end of the second symbol of the μ numerology frame. The following recursive equation is obtained:

$$N_{CP0}(\mu) = 2N_{CP0}(\mu - 1) - N_{CP} \tag{2.23}$$

The time duration of the first CP, T_{CP0}, immediately follows:

$$T_{CP0} = f_s \cdot N_{CP0} \tag{2.24}$$

2.3.3 5G Resource Grid and Bandwidth

In NR, a resource grid is defined as in LTE (Figure 2.11).

One PRB still consists of 12 subcarriers so that the PRB bandwidth is:

$$BW_{PRB} = 12 \cdot \Delta f \tag{2.25}$$

where Δf is given by (2.16). Minimum and maximum number of PRBs depends on the numerology as specified in Table 2.3.

The number of occupied subcarriers can be obtained multiplying the figures in Table 2.3 by 12: its maximum value is 275x12=3,300. Based on Table 2.3 and (2.25), maximum and minimum bandwidth, BW_{min} and BW_{max}, can be calculated at different numerologies:

$$BW_{min,max} = N_{PRBS,\ min,\ max} \cdot BW_{PRB} \tag{2.26}$$

The values of BW_{min} and BW_{max} are reported in Table 2.4.

Table 2.4 shows that NR has a better spectral efficiency than LTE. For example, with μ=4, NR can use 99% of an available BW of 100 MHz, while only 90% (see Table 2.2) is achievable frequency multiplexing five 20 MHz LTE signals.

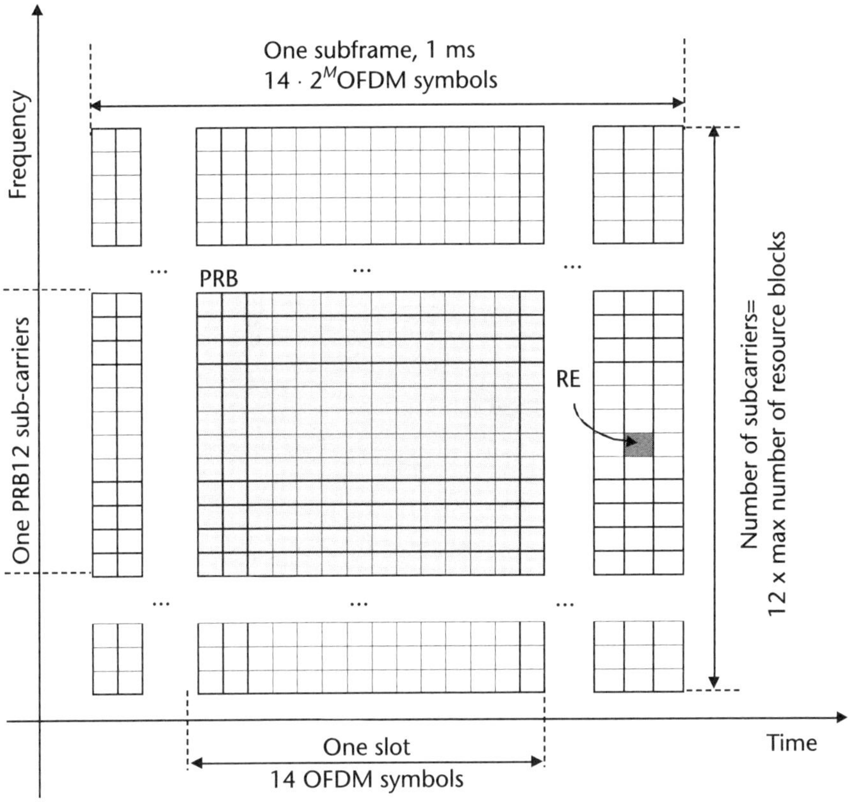

Figure 2.11 NR resource grid.

Table 2.3 Minimum and Maximum Number of PRBs in NR

μ	*Min Number of PRBs*, $N_{PRBS,\ min}$	*Max Number of PRBs*, $N_{PRBS,\ max}$
0	24	275
1	24	275
2	24	275
3	24	275
4	24	138

Table 2.4 Minimum and Maximum NR Bandwidth for Different Numerologies

μ	BW_{min}	BW_{max}
0	4.32 MHz	49.5 MHz
1	8.64 MHz	99.0 MHz
2	17.28 MHz	198.0 MHz
3	34.56 MHz	396.0 MHz
4	69.12 MHz	397.44 MHz

2.3.4 Time Division Duplex 5G Systems

The concept of TDD subframe is similar in NR and in LTE (see Section 2.2.5), with a substantial difference: In LTE, time intervals can be assigned to the UL or the DL with the granularity of a subframe, whereas NR allows the assignment of individual symbols, as shown in Figure 2.5. This generates many possible slot configurations, much more than in LTE, as specified in [2]. TDD systems find applications not only in the traditional mobile communications between UE and base station, but also in new scenarios like device-to-device communications, where the DL slot is used by the device that starts or schedules the transmission, and the UL slot is used by the device that responds to the request. A NR slot can be all DL, all UL, and mixed DL/UL. Three examples of different types of slot are shown in Figure 2.12 [14]: Figure 2.12(a) shows a mixed DL/UL slot with heavy DL transmission; Figure 2.12(b) shows a mixed DL/UL slot with heavy UL transmission and one symbol used for DL control; Figure 2.12(c) shows an all DL slot with a late start after the GB.

In TDD systems, the UE is informed whether a symbol is allocated to the downlink, uplink or is flexible (e.g., usable for both) by means of a field in the frame, called slot format indication (SFI), configured by the radio resource control (RRC) system. Due to this purpose, the SFI carries an index to a preconfigured table in the UE. NR also envisages the use of mini slots to support transmissions starting at any time in the slot and having a duration shorter than the duration of a regular slot (Figure 2.13). A mini slot can be as short as one OFDM symbol.

The mini slots are useful in various scenarios, as low-latency transmissions, transmissions in the unlicensed spectrum, and transmission in the millimeter waves (mm-Wave) spectrum. Low-latency scenarios require that the transmission begins suddenly, without waiting for the start of the next slot. In the unlicensed spectrum, it could be beneficial to start the transmission immediately after the listen-before-talk (LBT) procedure. LBT is a technique whereby a radio device senses its radio environment before starting the transmission. It is used to find a network where the device is allowed to operate or to find a free radio channel. Finally, in the mm-Wave spectrum, the available BW is so large that a few OFDM symbols are enough to carry a payload packet. NR supports the aggregation of consecutive slots. In Figure 2.14, two examples of aggregation of two slots are reported, for heavy DL and UL transmission.

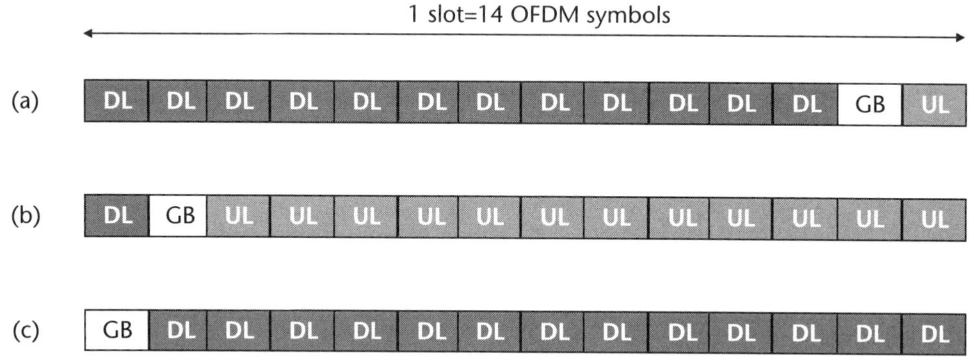

Figure 2.12 Examples of TDD slot configurations in NR.

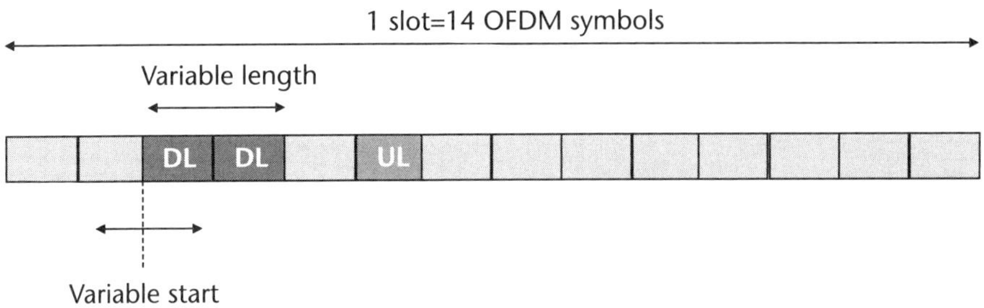

Figure 2.13 Concept of mini slot in NR.

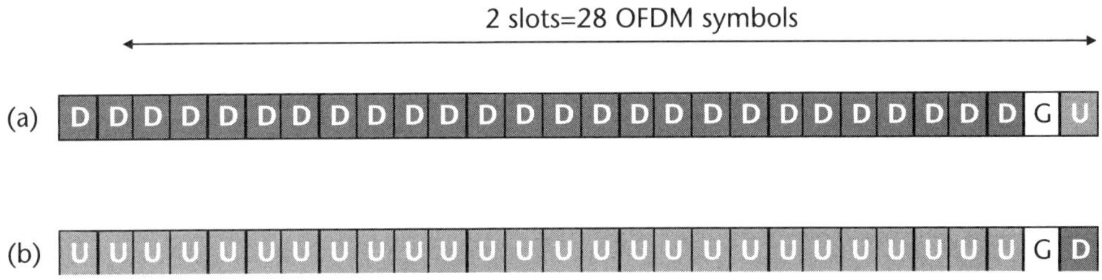

Figure 2.14 Slot aggregation in NR: (a) Heavy DL transmission, (b) heavy UL transmission.

The slot aggregation is useful for services that do not require extremely low latency, where a longer transmission time helps to reduce the overhead, due to the switchover between UL and DL, or to the transmission of reference and control signals. Frame structure and mechanisms illustrated for TDD systems apply also to FDD systems, enabling simultaneous reception and transmission, with DL and UL overlapping in time.

2.3.5 5G Physical Layer Parameters

The main physical layer parameters of the NR standard are summarized in Tables 2.5 and 2.6. All the values that depend on the FFT size are calculated with N=2,048.

- Frame duration: 10 ms;
- Subframe duration: 1 ms;
- OFDM symbols per slot: 14;
- Minimum number of PRBs: 24;

Table 2.5 NR Time Frame Parameters

μ	f_s	*Slots per Subframe*	*Slots per Frame*	*Slot Duration*	*Symbol Duration (No CP)*	*CP Duration*	*Length of First CP in 0.5 ms*	*Duration of First CP in 0.5 ms*
0	30.72 MHz	1	10	1 ms	66.67 μs	4.69 μs	160	5.21 μs
1	61.44 MHz	2	20	0.5 ms	33.33 μs	2.34 μs	176	2.87 μs
2	122.88 MHz	4	40	250 μs	16.67 μs	1.17 μs	208	1.69 μs
3	245.76 MHz	8	80	125 μs	8.33 μs	0.59 μs	272	1.11 μs
4	491.52 MHz	16	160	67.5 μs	4.17 μs	293.12 ns	400	0.81 μs

Table 2.6 NR Bandwidth Parameters

μ	Δf	BW_{PRB}	*Max Number of PRBs*	$BWmin$	$BWmax$
0	15 kHz	180 kHz	275	4.32 MHz	49.5 MHz
1	30 kHz	360 kHz	275	8.64 MHz	99.0 MHz
2	60 kHz	720 kHz	275	17.28 MHz	198.0 MHz
3	120 kHz	1.44 MHz	275	34.56 MHz	396.0 MHz
4	240 kHz	2.88 MHz	138	69.12 MHz	397.44 MHz

- Subcarriers per PRB: 12;
- Number of samples per CP, expect the first CP in 0.5 ms: 144.

2.4 Multiple Antenna Systems and Beamforming

A multiple antenna (or MIMO) system is a radio communication system with M transmitting antennas and N receiving antennas, shortened to MxN MIMO. In the 2x2 MIMO system of Figure 2.15, the received signals r_0 and r_1 are combinations of two transmitted signals, x_0 and x_1, called layers.

The received signals are different because the propagation conditions are different for the four channel paths indicated in Figure 2.15. Indeed, each path includes not only the line-of-sight (LOS) direct path but also multiple paths created by reflection, diffraction and scattering from the surrounding environment. Due to propagation impairments, some antenna of a MIMO system may not be able to correctly recover the transmitted data stream. However, this may become possible by using a technique known as precoding, which consists in the generation of new transmitted waveforms (y_0 and y_1 in Figure 2.16) obtained by means of a linear combination of the layers, according to (2.27).

$$\underline{y} = W\underline{x} \tag{2.27}$$

In (2.27), $\underline{x}$ and $\underline{y}$ are arrays of layers and precoded signals, respectively, and W is an MxN matrix that weights each path with a complex number. The weights

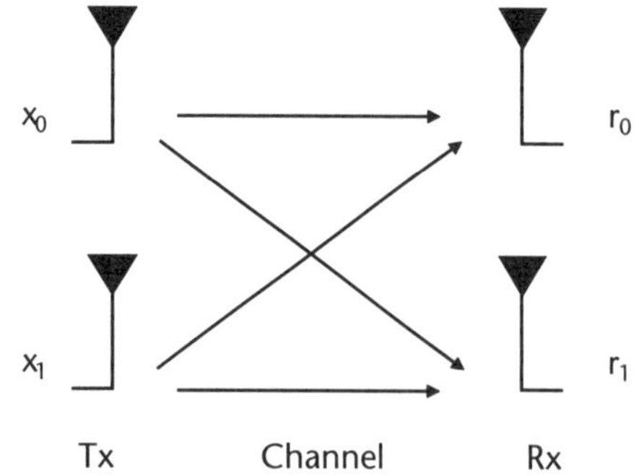

Figure 2.15 MIMO concept.

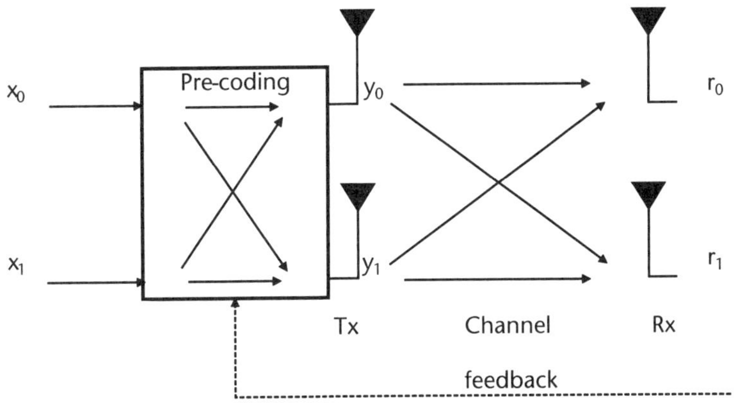

Figure 2.16 MIMO systems with precoding.

are calculated with the help of a feedback signal from receiver (Rx) to transmitter (Tx) so that, based on the received signals, the transmitter can estimate the channel response and calculate the weights on each path.

MIMO systems enable the use of techniques, such as spatial multiplexing, beamforming, and Rx or Tx diversity. Spatial multiplexing is used to increase the transmission data rate by sending independent data signals, sometimes referred as data streams, from multiple transmitting antennas to the same receiving antenna. With Tx diversity, a receiving antenna receives copies of the same signal, transmitted from multiple antennas and propagating on independent paths in such a way that the probability that all signals fade simultaneously is significantly reduced. A similar principle underlies the Rx diversity, where a signal transmitted by a single antenna is sent to multiple receiving antennas. Finally, beamforming is the ability to transmit the energy of a radio signal towards a specific receiver direction, or, when receiving, to collect signal energy from a specific transmission direction. The needed directional antenna is realized by feeding the different elements of an array of antennas with phase-shifted copies of the same signal, so that constructive interference occurs in only one direction, so to increasing received signal strength and end-user throughput.

The transmission data rate of a MIMO system increases with the number of layers. Indeed, the same time and frequency resources (i.e., the PRBs in the resource grid) can be reused at different layers up to a number of times equal to the available antenna ports (channel rank). This is exemplified in Figure 2.17, which makes evident the meaning of the term layer, intended as a third spatial dimension of the resource grid, in addition to time and frequency.

MIMO systems were already envisaged in LTE, but it is only with the advent of 5G that their potentiality is fully exploited, due to the introduction of large size antenna arrays (massive MIMO), facilitated by operation at high radio frequencies, which allows the use of smaller antenna elements. A typical example of antenna array used in 5G consists of 64 radiating elements, arranged in 8 rows and 8 columns, forming a square array that, in the mm-Wave frequency range (above 30 GHz) is no wider than a few square inches. Increasing the number of radiating elements allows for increasing the number of layers to improve the beamforming accuracy, generating very focused narrow beams, or, partitioning the antenna array in sub-arrays, to

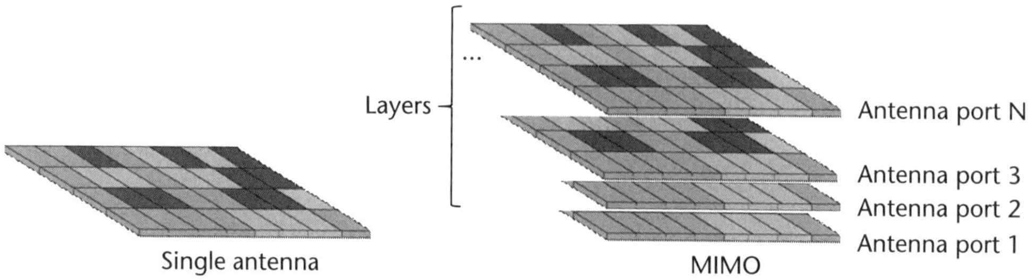

Figure 2.17 Reuse of time and frequency resources (PRB) in MIMO systems.

generate multiple beams pointing to different users. Relevant use cases are reported in [15]. For example, higher user capacity can be achieved transmitting multiple layers to a single user (single-user MIMO) or different layers can be simultaneously sent, in separate beams, to different users (multiuser MIMO), over the same time and frequency resources. Dynamically adjusting gains and phase offsets of the antenna elements of an array, it is possible to follow the user as they move within the cell, always sending information to the device instead of broadcasting it across the entire cell. All these techniques increase the network capacity and improve the end-user experience, especially in congested scenarios, such as densely populated cities, where there is no available free frequency band and the interference severely affects the quality of the radio channel.

2.5 Signal Processing Chain in 5G

A detailed description of the various types of channels defined by the NR standard for user data, control data and synchronization information is out of the scope of this book. The interested reader can find it in [2]. Each type of channel has its specific frame format and processing chain. Moreover, nonstandard and implementation-specific functionalities may be adopted by different equipment vendors. The purpose of this section is to describe the main functionalities, illustrated in the signal processing chain [16] of Figure 2.18, that are expected to be common to all type of channels. In the remainder of this section, the downlink chain at the top of Figure 2.18 will be taken as reference. Italic font will used through this section to highlight the link between the text and the processing blocks in Figure 2.18.

The *data bits* are sent in units called transport blocks (TBs) to which a cyclic redundancy code (CRC) is attached. The CRC is 24 bits for a TB larger than 3,824 bits, and 16 bits in all other cases. The TBs can be segmented in multiple code blocks: When this happens, each code block has its own CRC. Then, the transport blocks are *coded* to perform forward error correction (FEC). NR uses low-density parity check (LDPC) codes for the data channels and polar codes for the control channels. LDPC codes are a class of linear block codes, which are defined by means of a parity check matrix. In the parity check matrix, the number of columns equal the number of coded bits and each row contains the coefficients of a parity check equation. The matrix of an LDPC code contains only a few 1s in comparison to the amount of 0s, hence their name. The LDPC codes were introduced in 1960 by Gallager, but at that time they were considered too complex to implement. Their

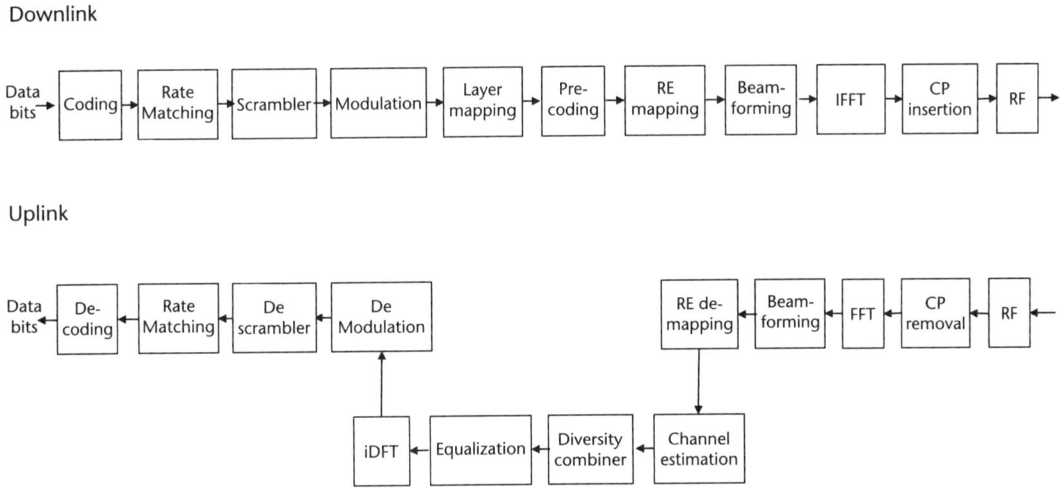

Figure 2.18 NR signal processing chain.

main advantage is that they can almost achieve the theoretical channel capacity with a decoding time that scales linearly with the codeword length. The LDPC codes used in NR are known as quasi-cyclic rate-compatible code. A detailed description can be found in [17]; here it is just worthwhile to mention that they can be implemented by means of parallel DSP. NR uses rate compatible codes, whose rate, defined as the ratio of number of information bits and number of codeword bits, can be adapted according to the available channel state information (CSI). An effective way to realize a rate compatible code is to puncture a base code, informing the receiver about the locations of the punctured symbols [18]. Since the decoder for the code with the lowest rate base is compatible with the decoders at higher code rates obtained by puncturing, no additional complexity is needed for achieving the adaptability. The structure of a parity check matrix is illustrated in Figure 2.19. The light gray part (at the top left of Figure 2.19) represents the matrix of the high rate base code. It is a systematic code, that is, an error correcting code where the input data is embedded in the encoded output (codeword). Usual rate values are 2/3 or 8/9. Additional parity bits are generated by extending the base matrix and including the rows and columns marked in dark grey (at the left bottom of Figure 2.19). Note that the parity-check matrix is smaller for higher code rates, as well as decoding complexity and latency. LDPC codes are usually decoded by using iterative algorithms, a technique known as soft decoding.

The Polar codes, which are also used in NR for control signaling, will not be described here. Introduced by Arıkan in 2008 [19] as the LDPC codes, they are a family of linear error correcting block codes that can get close to the Shannon capacity limit with reasonable decoding complexity. After coding, *rate matching* is a function that adjusts the number of coded bits to fit the resources available for data transmission. The available resources depend on the capacity used for other purposes, including reference signals, system information, control channels, and reserved resources. The coded bits selected in the rate-matching process are chosen from a circular buffer that the FEC encoder output is written into. After rate matching, bit-level interleaving is applied to each code block. More details can be found in [20]. After rate matching, the coded bit sequences are scrambled.

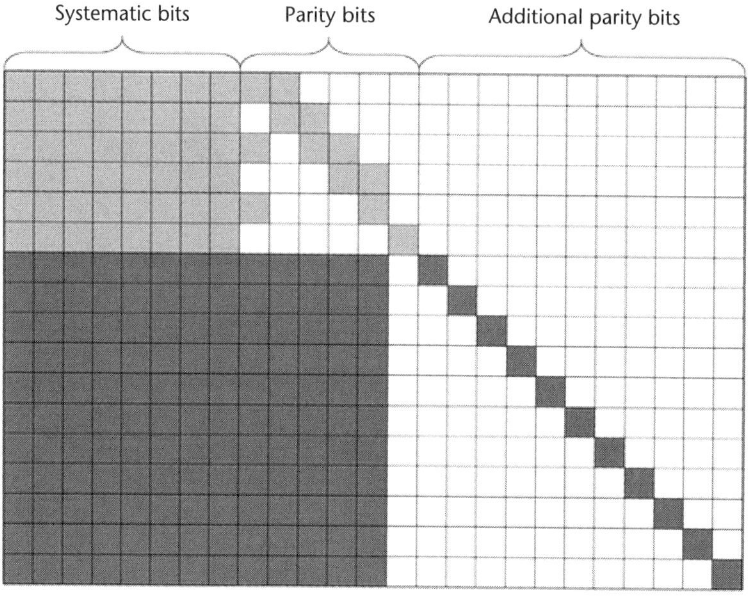

Figure 2.19 Structure of a parity check matrix. (Source: [14].)

Scrambling [21] is a digital processing technique that eliminates long strings of equal bits in a sequence, to help the synchronization at the receiver, eliminating the DC component and periodic bit patterns. Scrambling is usually performed sending a sequence of bits $\{b_{-n}, b_{-(n-1)}, \ldots, b_0\}$ to a shift register with *n* cells. Each element of the register is weighted with a binary digit (that is, bits weighted by 1 are selected and bits weighted by 0 are ignored) and, finally, all weighted bits (or, equivalently, the selected ones) are modulo-2 added. An example of scrambler with 7 cells is shown in Figure 2.20, which also reports a polynomial, whose meaning is evident from the figure, univocally associated to the scrambling code. NR uses Gold scrambling codes [22] of length, *n*, equal to 31.

After the scrambler, the signal is *modulated* (Section 2.3.1) and the modulation symbols are *mapped into the different layers* of a MIMO system (Section 2.4), which are *precoded* for the transmission (Section 2.4). The precoded modulation symbols are *mapped in the REs* of the OFDMA resource grid (Section 2.3.3). The *beamforming* block in Figure 2.18 sends phase-shifted copies of the input signals to the different elements of the Tx antenna array, an operation known as port

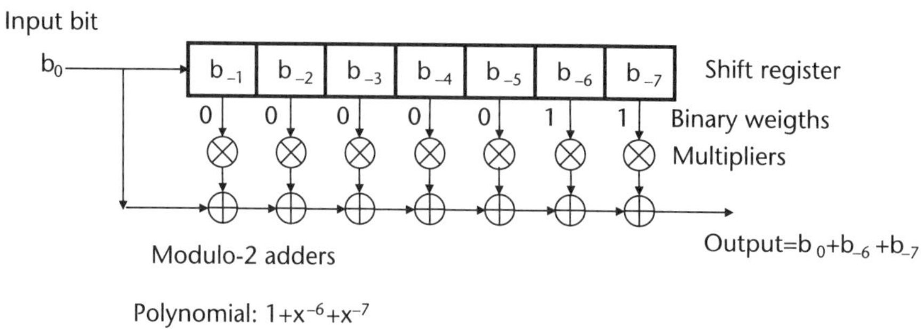

Figure 2.20 Example of scrambler.

expansion. Then, the *IFFT* block generate the OFDM signal (Section 2.2.1) and the CP is added (Figure 2.10). Finally, the signal is digital to analog converted and its frequency is up-converted to the radio frequency (*RF*).

The inverse sequence of operation is performed in the uplink (bottom of Figure 2.18), where additional blocks are present for *channel estimation and equalization*, and inverse discrete Fourier transform (*iDFT*), if DFTS-OFDM is used (Section 2.2.2). The *diversity combiner* block in the uplink inverts the precoding operation in the downlink.

2.6 Conclusions

This chapter illustrated the techniques used in the modern 4G and 5G mobile communication systems to generate the signal on-air. The operation of an OFDMA mobile system has been explained, together with the rules to calculate the main physical layer parameters, like symbol period, sampling frequency, CP duration, resource block size, and signal BW. The frame format in the two cases of FDD and TDD was also presented. Having discussed the physical layer, the next chapter will illustrate the upper layers of the radio protocol stack. Then, it will explain how the combination of physical layer specifications, such as the capacity, and upper layer requirements, such as the latency, affect the design of the radio access network architecture.

References

[1] "Ericsson Mobility Report 2018," https://www.ericsson.com/en/mobility-report.

[2] "NR Physical Channel and Modulation," 3GPP Technical Specification 38.211.

[3] Proakis, J. G., *Digital Communications*, MA: McGraw-Hill, 1995, pp. 163–173.

[4] Carlson, A. B., P. B. Crilly, and J. C. Rutledge, *Communication Systems*, MA: McGraw-Hill, 2002, pp. 144–147.

[5] Carlson A. B., P. B. Crilly, and J. C. Rutledge, *Communication Systems*, MA: McGraw-Hill, 2002, pp. 644–655.

[6] Seybold, J. S., *Introduction to RF Propagation*, Hoboken, NJ: John Wiley & Sons, 2005, pp. 163–206.

[7] Proakis, J. G., *Digital Communications*, MA: McGraw-Hill, 1995, pp. 636–680.

[8] Oppenheim, A. V., and R. W. Schafer, *Discrete-Time Signal Processing,* MA: Prentice Hall, 1989.

[9] Van Waterschoot, T., V. Le Nir, and J. Duplicy, "Analytical Expressions for the Power Spectral Density of CP-OFDM and ZP-OFDM Signals," *Signal Processing Letters*, Vol. 17, No. 4, April 2010.

[10] Hamza, A. S., S. S. Khalifa, and H. S. Hamza, "A Survey on Inter-Cell Interference Coordination Techniques in OFDMA-Based Cellular Networks," *Communications Surveys & Tutorials*, Vol. 15, No. 4, Fourth Quarter 2013.

[11] Van Nee, R., and A. De Wild, "Reducing the Peak-to-Average Power Ratio of OFDM," *Proc. VTC '98. 48th IEEE Vehicular Technology Conference. Pathway to Global Wireless Revolution.*

[12] Patenaude, F., and M. L. Moher, "A New Symbol Time Tracking Algorithm for p/2-BPSK and p/4-QPSK Modulations," *Proc. SUPERCOMM/ICC '92 Discovering a New World of Communications*, Vol. 3, 1992, pp. 1588–1592.

[13] Zaidi, A. A., et al., "OFDM Numerology Design for 5G New Radio to Support IoT, eMBB, and MBSFN," *Communications Standard Magazine*, Vol. 2, No. 2, June 2018.

[14] Zaidi, A. A., et al., "Designing for the Future: the 5G NR Physical Layer," Ericsson Technology Review, July 2017.

[15] Von Butovitsch, P., et al., "Advanced Antenna Systems for 5G Networks," Ericsson White Paper, November 2018.

[16] "eCPRI 1.2 presentation," http://www.cpri.info/spec.html.

[17] Myung, S., K. Yang, and J. Kim, "Quasi-Cyclic LDPC Codes for Fast Encoding," *IEEE Transactions on Information Theory,* Vol. 51, No. 8, August 2005.

[18] Ha, J., J. Kim, and D. Klinc, "Rate-Compatible Puncturing of Low-Density Parity-Check Codes," *Transactions on Information Theory*, Vol. 50, No. 11, November 2004.

[19] Arıkan, E., "Channel Polarization: A Method for Constructing Capacity-Achieving Codes for Symmetric Binary Input Memoryless Channels," *Transactions on Information Theory*, Vol. 55, No. 7, July 2009.

[20] Bertenyi, B., et al., "5G NR Radio Interface," *Journal of ICT, Combined Special Issue* 1 & 2, River Publishers, 2018.

[21] Carlson, A. B., and P. B. Crilly, J. C. Rutledge, *Communication Systems*, MA: Mc-Graw Hill, 2002, pp. 479–484.

[22] Dinan, E.H., and B. Jabbari, "Spreading Codes For Direct Sequence CDMA and Wideband CDMA Cellular Networks," *Communications Magazine*, Vol. 36, No. 9, September 1998.

CHAPTER 3

Radio Access Network Architecture

3.1 Introduction

The advent of 5G systems has led to major changes not only in the wireless access network, but in the whole network infrastructure—including the optical transport infrastructure. In a communication network, the transport domain oversees the connectivity between remote sites that perform different functionalities, such as traffic aggregation and routing. A transport network is conventionally partitioned into different segments, identified by features such as performed functionality, typical link distance, and aggregate capacity, but there is no universally agreed definition for these segments. A coarse orientation map is provided in Figure 3.1, which shows at its bottom the distance order of magnitude for the various segments.

The data traffic generated by user devices (mobile phones, laptops, smart TVs, etc.) is collected by mobile and fixed access networks through an air interface, wired optical fiber, or copper links, respectively. An access network is connected—usually via optical fiber—to an aggregation network, which hosts functionalities like packet switching and baseband processing of mobile traffic. An edge router connects the aggregation network to a metro network, where data is transported over high-capacity optical channels based on wavelength division multiplexing (WDM) and time division multiplexing (TDM) techniques such as the optical transport network (OTN) standard, defined by the Standardization Sector of the International Telecommunication Union (ITU-T)[1]. Hereinafter, the access and aggregation networks for mobile traffic will be considered as a single segment, referred to as 5G transport network or radio access network (RAN), as shown in Figure 3.1. In the RAN, the term backhaul (BH) is used for the link connecting a base station (BS) to a packet network, while fronthaul (FH) is the term used for links connecting a radio unit (RU)—equipment that performs a subset of low-level functions of the radio protocol stack—to a baseband processing and control unit that hosts upper-layer functionalities.

This chapter will first illustrate the new services enabled by 5G and their requirements. Then, the functionalities of the layers of the radio protocols stack will be described. Finally, it will be discussed how these functionalities can be partitioned among the different nodes of the transport network to satisfy the service requirements.

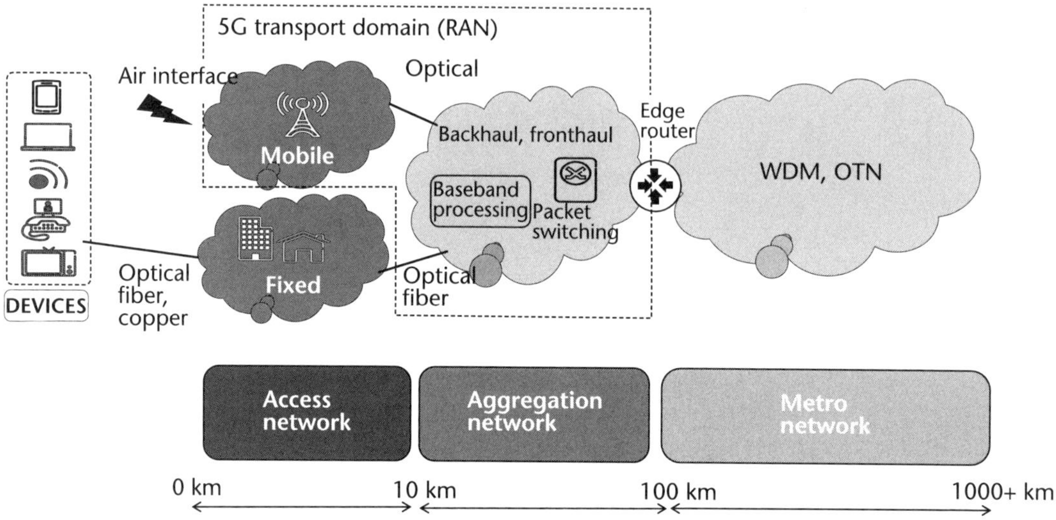

Figure 3.1 Network segments.

3.2 5G Use Cases and Requirements

The 5G mobile generation enables several new use cases in various industry and society sectors, such as automotive, construction, energy, health, manufacturing, and media [2]. Automated vehicle control for autonomous driving, remote medical surgery, and high-efficiency factories based on cloud robotics are just three examples on how 5G can change our lives [3]. Table 3.1 reports various 5G use cases and related requirements. Such a variety of scenarios makes the design of a single transport network for all 5G applications more difficult now than in the past.

To better deal with this complexity, the 5G use cases have been classified into three wide classes, also designed as 5G services: mMTC, otherwise known as massive IoT; URLLC, otherwise known as critical Machine Type Communication (cMTC), or critical IoT; and eMBB.

mMTC is designed for applications, such as monitoring of buildings and infrastructure, smart agriculture, and logistics and tracking, and is characterized by wide area coverage and high device density (~100,000 devices/km^2). mMTC applications tend to be quite delay-tolerant. Since the mMTC devices are typically battery powered, connectivity must be provided with high energy efficiency. This implies the use of low complexity software and hardware, small data payloads, and devices capable of staying inactive for long periods of time.

Examples of URLLC applications include the automated energy distribution in a smart grid, and the control of industrial processes where the communication must occur in real time. URLLC applications require high reliability, low latency within the order of one millisecond, and a high level of security.

eMBB envisages all the use cases requiring extremely high data rates and low latency, and it also offers wide coverage. Connectivity and bandwidth are quite uniform over the coverage area, and the performance degrades gradually as the number of users increases.

To understand how the RAN architecture is impacted by the requirements of the 5G services, it is necessary to have some knowledge about the radio protocol

Table 3.1 Requirements for Different 5G Use Cases

Use Cases	*Requirements*	*Desired Value*
Autonomous vehicle control	Latency	5 ms
	Availability	99.999%
	Reliability	99.999%
Emergency communication	Availability	99.9% victim discovery rate
	Energy efficiency	One-week battery life
Factory cell automation	Latency	Down to below 1 ms
	Reliability	Down to packet loss of less than 10^{-9}
High-speed train	Traffic density	DL: 100 Gbps/km^2; UL: 50 Gbps/km^2
	User throughput	DL: 50 Mbps, UL: 25 Mbps
	Mobility	500 kmph
	Latency	10 ms
Large outdoor event	User throughput	30 Mbps
	Traffic density	900 Gbps/km^2
	Connection density	Four devices/m^2
	Reliability	Outage probability < 1%
Massive numbers of geographically dispersed devices	Connection density	1,000,000 devices/km^2
	Availability	99.9% coverage
	Energy efficiency	10-year battery life
Media on demand	User throughput	15 Mbps
	Latency	5s (start application), 200 ms (after link interruptions)
	Connection density	4,000 devices/km^2
	Traffic density	60 Gbps/km^2
	Availability	95% coverage
Remote surgery and examination	Latency	Down to 1 ms
	Reliability	99.999%
Shopping mall	User throughput	DL: 300 Mbps UL: 60 Mbps
	Availability	95% for all applications, and 99% for safety-related applications
	Reliability	95% for all applications, and 99% for safety-related applications
Smart city	User throughput	DL: 300 Mbps, UL: 60 Mbps
	Traffic density	700 Gbps/km^2
	Connection density	200,000 devices/km^2
Stadium	User throughput	0.3-20 Mbps
	Traffic density	0.1-10 Mbps/m^2
Tele-protection in smart grid network	Latency	8ms
	Reliability	99.999%
Traffic jam	Traffic density	480 Gbps/km^2
	User throughput	DL: 100 Mbps, UL: 20 Mbps
	Availability	95%
Virtual and augmented reality	User throughput	4-28 Gbps
	Latency	< 7ms
Broadband to the home	Connection density	4,000 devices/km^2
	Traffic density	60Gbps/km^2

Source: [3].

stack, which will be described in the next section. However, it is already evident from the traffic density and the user throughput required by some of the use cases in Table 3.1 (which can achieve several hundreds of Gbit/s per Km^2 and hundreds of Mbit/s, respectively) that a transport network for 5G will have to accommodate capacities much higher than the typical capacity of a current aggregation network. Moreover, for services requiring an end-to-end latency of some milliseconds, the delay introduced by the transport network cannot be higher than tens or hundreds of microseconds.

3.3 The Radio Protocol Stack

The radio protocol stack is illustrated at the left side of Figure 3.2. The stack is similar for the LTE and NR standards. It consists of physical layer (PHY), medium access control (MAC), radio link control (RLC), Packet Data Convergence Protocol (PDCP), and radio resource control (RRC). The stack is slightly different for the user plane and the control plane: PHY, MAC, RLC and PDCP are defined for both user and control plane data, while the RRC layer lies only in the control plane. Moreover, in 5G, a new user plane layer has been added above the PDCP, named the Service Data Adaptation Protocol (SDAP); this is not shown in Figure 3.2 for the sake of simplicity.

PHY is a physical layer, or layer 1, protocol; MAC, RLC and PDCP are data link, or layer 2, protocols; and the RRC is a network, or layer 3, protocol. Reading it from bottom to top, the right side of Figure 3.2 shows how Internet Protocol (IP) packets are obtained from PHY transport blocks (TBs). The main protocol functions are listed in the following, without demanding completeness.

PHY adapts to the air interface, described in Chapter 2, the MAC protocol data units (PDUs), and vice versa. Its main responsibility is the link adaptation, or adaptive coding and modulation (ACM), denoting any operation to match modulation,

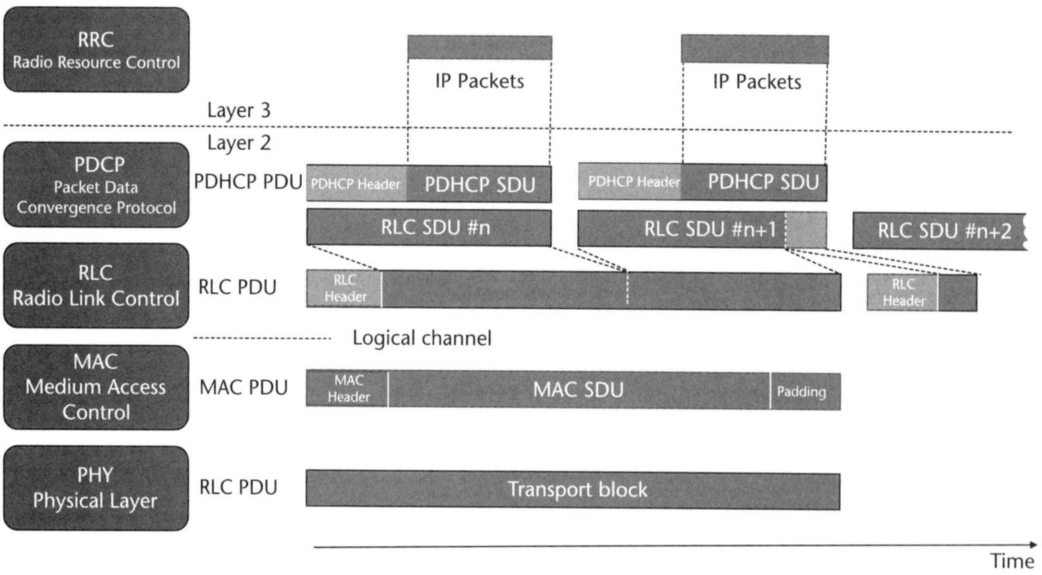

Figure 3.2 Radio protocol stack.

coding and other air interface features to the quality of the radio link. Other PHY functions are the control of the transmitted power and the cell search during the initial synchronization procedure and the handover. Moreover, PHY measures the quality of the radio link, passing that information to the control plane.

The MAC transport channel is arranged in service data units (SDUs). In a MAC SDU, RLC logical channels are mapped into PHY TBs and vice versa. A TB delivered from the PHY to the MAC contains the data of the previous subframe and can carry multiple or partial packets, depending on the used scheduling and modulation schemes. The MAC adds to the TB overhead and padding bits. Moreover, it breaks down different logical channels out of the TB, providing them to the layers above, a function known as logical mapping. The MAC is also responsible for: Reporting scheduling information; performing error correction through the hybrid automatic repeat request (HARQ) protocol (see Section 3.4); dynamic scheduling for priority handling among UEs; and priority handling between the logical channels of the same UE.

The RLC is responsible for the transfer of PDUs to and from upper layers. It segments one RLC PDU into multiple RLC SDUs, or portions of them. To ensure the in-order delivery of the SDUs, a sequence number is added to the RLC header; it is independent of the SDU sequence number, which is the responsibility of the PDCP. Error correction is performed through automatic repeat request (ARQ) retransmission (see Section 3.4) of RLC SDUs (while the HARQ in the MAC acts on the TBs). The RLC can be configured in three modes: Transparent mode (TM), acknowledged mode (AM), and unacknowledged mode (UM). The TM is used for control plane RLC messages during the initial connection; it uses no header, but simply passes the message through. UM and AM modes indicate in the RLC header whether ARQ retransmission is performed or not. The RLC can also discard RLC SDUs. The far-right side of Figure 3.2 shows the concatenation of two RLC SDUs in a single SDU, an operation that is possible in LTE, but in 5G has been moved to the MAC.

The responsibilities of the PDCP include: Header compression and decompression of IP packets to save bandwidth on air; sequence numbering and in-sequence delivery of upper layer PDUs; duplicate detection and removal; retransmission, discard or duplication of PDCP SDUs; data ciphering and deciphering; and integrity verification and protection of control plane data.

The RRC is the first layer 3 protocol in the stack. It broadcasts system information about the access stratum (AS) and the nonaccess stratum (NAS), which are IP protocols above the RRC, not shown in Figure 3.2. The NAS protocols support the mobility of the UE and the session management procedures to establish and maintain the IP connectivity between UE and packet data network gateway. The RRC is also responsible for: Paging (i.e., waking up an UE when there is some data for it); establishment, maintenance and release of an RRC connection between the UE and the network, including addition, modification, and release of carrier aggregation or dual connectivity; security functions, including key management; mobility functions, including handover and context transfer; UE cell selection; quality of service (QoS) management functions; reporting and control of the link measurements made by the UE; detection of and recovery of radio link failures; and establishment, configuration, maintenance and release of point-to-point radio bearers (RBs). An RB specifies the layer 2 and layer 1 configurations that guarantee a certain QoS between two points of the network. An RB can also be considered as a logical channel offered by the layer 2 to higher layers for the transfer of user and control data. In other words, an RB is a service access point (SAP) between layer 2 and the upper layers.

Finally, it has already been mentioned that in 5G a new user plane layer 3 protocol, SDAP, has been included. Its main functions are the mapping between QoS flows and data RBs, and marking QoS flow identifiers (IDs) in DL and UL packets.

3.4 The HARQ Protocol

The HARQ is one of the most important functions in a mobile communication system. It is performed by the MAC in cooperation with the PHY and consists of the retransmission of TBs when errors are detected at the receiver. The HARQ is an improvement of the ARQ protocol, which is a simpler retransmission scheme whose working principle is illustrated in Figure 3.3.

In Figure 3.3, a first transport block, TB 1, is transmitted. The receiver detects the absence of errors, (for example, checking that the CRC appended to the TB is correct) and notifies it to the transmitter by sending an acknowledgement (ACK) message, ACK 1, so that the transmitter can send a second TB. This time, the receiver detects error (for example, due to an impairment of the transmission channel), so it discards the TB and sends a nonacknowledgement (NACK) message, NACK 2, to the transmitter. The transmitter does a second attempt, which fails as well. Finally, after a third successful transmission attempt, a new transport block, TB 3 can be transmitted.

The main issue of the ARQ protocol is that the same TB is retransmitted until no errors are detected, blocking the transmission of the subsequent TBs. This unacceptably impairs the link latency, which is critical in some 5G applications (see Section 3.2). To mitigate this issue, the HARQ was introduced. In the HARQ, the different instances of a same failed TB are not discarded, but are rather retained in a buffer and combined so that the correct information can be retrieved from the partial information still present in the corrupted TBs. This technique is known as soft combining [4] and requires coded data (see Section 2.5); different version of the same code, for example obtained by puncturing, are used at each retransmission.

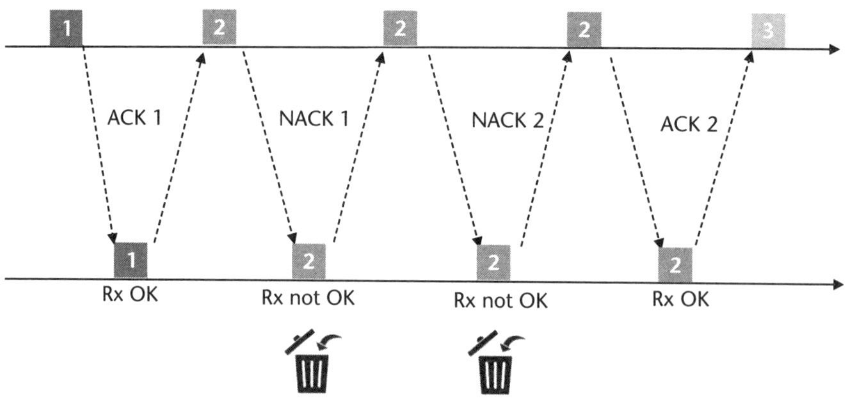

Figure 3.3 ARQ retransmission mechanism.

The HARQ uses two main soft combining methods: Chase combining and incremental redundancy. With chase combining, same data and parity bits are sent at every retransmission, and the receiver uses maximum-ratio combining techniques [5] to combine the received bits with the same bits from previous retransmissions. Using incremental redundancy, different coded bits are generated at each retransmission from the same set of information bits. The different sets of coded bits have a different redundancy. They are generated by puncturing the encoder output so that only a fraction of coded bits is transmitted. The puncturing pattern is different at each retransmission, so that comparing different codewords allows the receiver to recover additional information. Figure 3.4 shows how the HARQ works, with the same example used in Figure 3.3.

Differently coded versions of the same TB are indicated by the number in brackets. In this example, the soft decoding of two stored replicas of TB 2 is sufficient to correct the errors, so that TB 3 can be sent in advance, as compared to Figure 3.3.

A remarkable advantage of the HARQ, enabled by the use of buffers, is the possibility to run more than one process in parallel to retry several outstanding TBs. This further reduces the latency and allows for a continuous transmission, which cannot be achieved with a simple stop-and-wait scheme like the ARQ. When one HARQ process is waiting for an ACK, another process can use the channel to send data. The data processed by the different HARQ processes is then arranged by the MAC in logical channels and forwarded to the RLC for reordering.

More implementation variants of the HARQ exist. Type I HARQ exploits both CRC and FEC information added to TB prior to the transmission. When a coded data block is received, the receiver first tries to correct the transmission errors by using the FEC decoder. If the FEC is able to correct all the errors, an ACK is sent to the transmitter. If some error could not be corrected, the receiver acknowledges it by checking the CRC, the coded data block is rejected, and a retransmission is requested by sending a NACK. In Type II HARQ, the first transmission only contains data and CRC parity bits, like in the ARQ. If no error is detected, an ACK is sent. If errors are detected, a second retransmission is asked, containing both FEC and

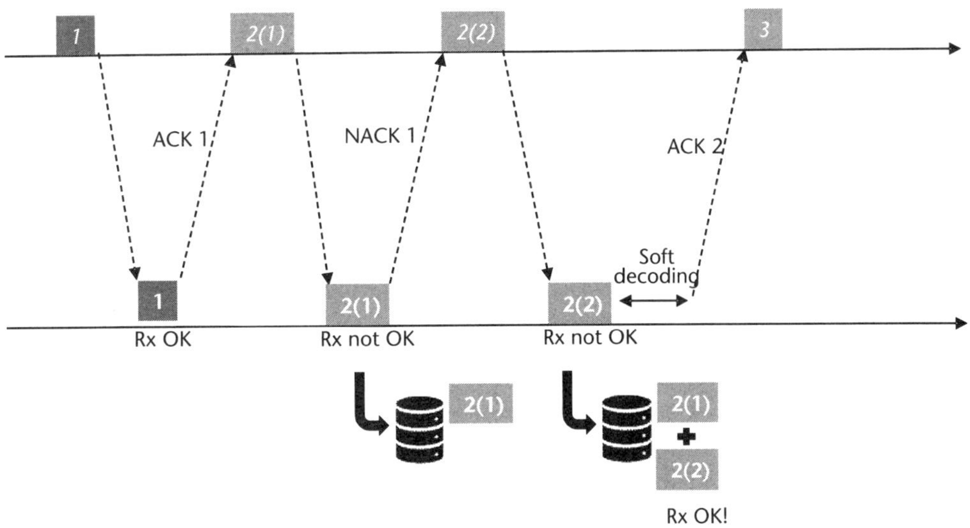

Figure 3.4 HARQ working principle.

CRC parity bits. Then, if the FEC decoder could correct all the errors, an ACK is sent to the transmitter. Otherwise, error correction is attempted by soft combining the information received from both the retransmissions. The amount of retransmitted data is significantly less in Type II HARQ compared with Type I HARQ because the number of CRC parity bits is typically a small fraction of the whole message length, whereas number of data and parity bits are often comparable in a FEC codeword.

The HARQ scheduling mechanisms in NR are outlined in the following; more details can be found in [6]. At the UE, the time intervals available for the retransmission are configured semi-statically by the RRC, separately for DL and UL. In both UL and DL, the HARQ is asynchronous and adaptive, which is especially useful in eMBB applications, while in LTE, the HARQ was synchronous and non-adaptive in the UL. Scheduling information to allow flexible timing is provided through fields in the DL control information (DCI), such as time between DL/UL assignment and corresponding DL/UL data transmission, and time between DL data reception and corresponding ACK. In NR, the UE manages a set of values which affect the minimum HARQ processing time, including delay between DL data reception and corresponding ACK transmission, and delay between UL grant reception and corresponding UL data transmission. Moreover, the UE is required to indicate to the BS its minimum HARQ processing time capability.

The retransmission and scheduling mechanisms discussed in this section are essential to any mobile communication system and absorb the most of its latency budget. In 5G, the latency is critical, especially for URLLC and eMBB services. The next section will analyze the latency budget and illustrate latency reduction strategies in a mobile communication system.

3.5 Latency Budget in Mobile Communication Systems

The concept of transmission time interval (TTI) was introduced in Section 2.2.2 as fundamental time unit in a radio link. Typically, but not always, one TTI equals the duration of a subframe; for example, in Figure 2.13, mini slots shorter than a subframe are used for data transmission. A more general definition of TTI is based on the fact that data is passed from the MAC to the PHY once per TTI. Of course, the higher the TTI, the higher the latency. The latency budget for the LTE user plane is reported in Annex B of [7]. As shown in Figure 3.5 for a TDD system, it is partitioned in UE processing time, eNodeB processing times, and TTI, which is 1 ms for LTE.

In Figure 3.5, t_{FA} is the time required for frame alignment. It depends on the frame structure (see Section 2.2.5), and in a TDD system, it varies from 0.6–1.7 ms in DL and 1.1–5 ms in UL.

The HARQ retransmissions introduce further delay, so that the total average one-way delay for the user plane can be estimated as:

$$Delay_{one\ way} = 3.5ms + t_{FA} + t_{HARQ\_RTT} \cdot P_{BLER} \tag{3.1}$$

where:

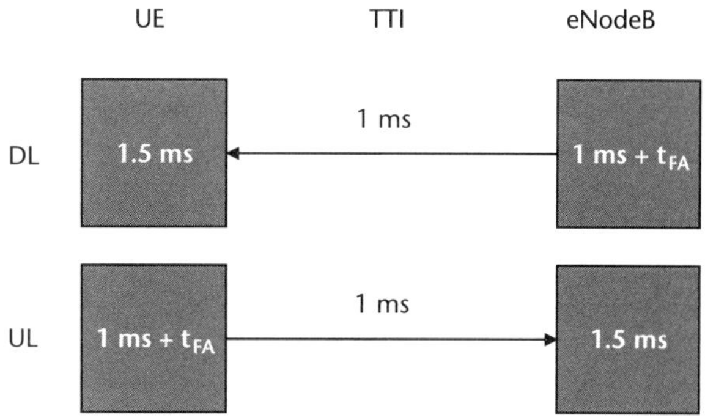

Figure 3.5 Latency budget in TDD LTE.

3.5 ms is the processing time at UE and eNodeB;

t_{FA} is the frame alignment time;

t_{HARQ_RTT} is the retransmission round-trip time (RTT). It depends on the TDD frame configuration, and spans from 9.8–12.4 ms in DL and 10–11.6 ms in UL;

P_{BLER} is the block error rate (BLER) probability, which is roughly equal to the ratio between the number of retransmitted and the total number of TBs. A typical assumption is P_{BLER}~10%.

Based on the above assumptions, in the best-case scenario, it is Delay $_{\text{one way}}$ ~5 ms. The exact values for the various TDD configurations can be found in the Annex B of [7].

In the UL, additional delay is due to the scheduling request (SR) sent by the UE to the network before sending data. This delay depends on scheduling protocol and system configuration; various cases are reported in the Annex B of [7]. Table 3.2 shows an example for a synchronized FDD transmission initiated by the UE, assuming 5 ms allocated to an SR. The resulting latency is 11.5 ms in error-free conditions, that is, no HARQ retransmission, which does not meet the latency requirement set by the 3GPP for URLLC services (0.5 ms in both DL and UL when no high reliability is required, or 1 ms with 10^{-5} reliability, referred to as a 32-byte packet).

Latency reduction strategies are already envisaged in LTE [8-9], but only in NR it is possible to dramatically reduce the TTI and all processing times proportional

Table 3.2 Example of Uplink Latency, Including Scheduling Delay

Description	*Time (ms)*
Average delay to next SR opportunity	2.5
UE sends scheduling request	1
eNodeB decodes scheduling request and generates the scheduling grant	3
Transmission of scheduling grant	1
UE processing delay (decoding of scheduling grant + PHY encoding of UL data)	3
Transmission of UL data	1
Total delay	11.5

to it [see (3.1) and Table 3.2]. In NR, the OFDM symbol duration depends on the numerology (see Section 2.3.2), so that for adopting a higher subcarrier spacing, the TTI is proportionally shortened. The other approach for reducing the latency is to allow the TTI to have a small number of OFDM symbols, for example, one or two. Various TTI lengths can be assigned flexibly to the UE, dependently on the service type. To allow the coexistence of UEs using different length TTIs, NR makes it possible to multiplex TTIs with a different number of OFDM symbols on the same carrier. To do so, the reference signals used for data demodulation are confined within a small number of OFDM symbols [8].

3.6 RAN Functional Split

3.6.1 Radio Split Architecture

In a radio split architecture, different layers of the protocol stack, or even portions of a same layer, are distributed in different nodes of the RAN instead of all residing in the same BS equipment. There are several reasons for such functional decomposition. Originally, it was thought to have leaner equipment at the antenna site, to improve power consumption and reliability, and simplify the installation practices. Furthermore, centralizing some functionalities in a single, central unit that serves mode distributed RUs can improve cell coordination, load management, and adaptive performance optimization. A specific use case is when layer 3 functionalities are moved in a datacenter, enabling the virtualization of the network functions (called network function virtualization [NFV]).

The functional decomposition of the BS has a high impact on the transport network. For example, splitting MAC and PHY into two different nodes of the network imposes stringent latency requirements to the link connecting the two nodes, which becomes part of the HARQ retransmission loop. If the link includes packet switches, care must be paid to limit packet queuing and processing times within the latency budget. Not only the latency, but also the capacity depends on the split point: Split options closer to the physical layer (low-level split) demand more bandwidth than split options closer to layer 3 (high-level split), where data is efficiently carried by IP packets and overhead from lower layers has been removed.

There is no optimal split for every situation, but the choice depends on network constraints and supported services, such as QoS settings of the offered services as low latency and high throughput; user density and traffic demand on the covered geographical area; and achievable capacity and latency of a preexisting transport network.

3.6.2 Functional Split Options

A detailed analysis of the various split options (Figure 3.6) is carried out in [10]. Each split option (indicated by an arrow in Figure 3.6) is associated with a number—higher numbers correspond to lower-level split options.

According to [10], for a given split option, central unit (CU) will indicate the equipment responsible for the high level functions, at the left side of the arrow, and distributed unit (DU) will indicate the equipment responsible for the low-level

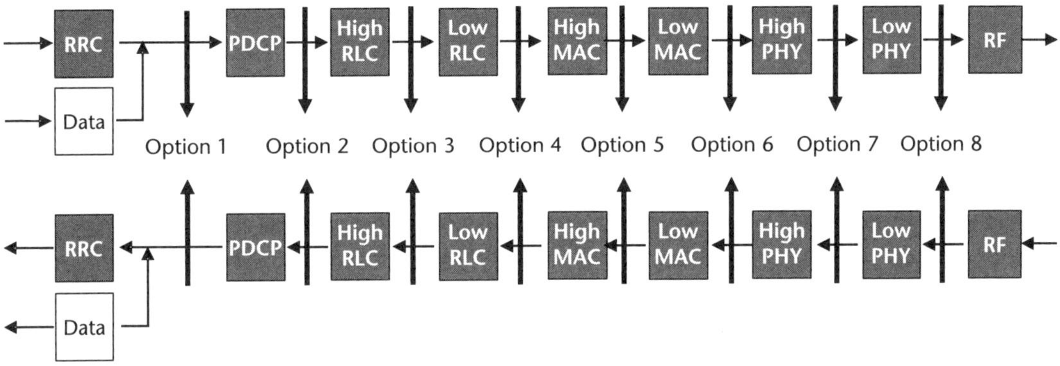

Figure 3.6 Functional split between a central and distributed unit.

functions, featured at the right side of the arrow. A summary is provided in Table 3.3 for the various options; note that intralayer split points are defined for RLC, MAC and PHY.

The main features of the various split options are outlined as follows:

Option 1 allows the separation of the user plane from a centralized control plane; for example, based on the Software-defined network (SDN) paradigm. Option 1 is beneficial when the user data need to be processed close to the transmission point, as in edge computing or low-latency services.

Option 2 allows the aggregation and the centralization of data traffic originating from NR and LTE networks. It has two variants, Option 2-1 and Option 2-2, referring to the case of PDCP and RRC hosted in the same CU, or split in two different CUs for the user plane and the control plane, respectively. In separating the user plane from the control plane, Option 2-2 enables a cost-effective network planning, where the number of network nodes scales with the user plane traffic load.

Table 3.3 Function Split Between CU and DU

	RRC	*PDCP*	*High RLC*	*Low RLC*	*High MAC*	*Low MAC*(1)	*High PHY*	*Low PHY*	*RF*
Option 1	CU	DU	DU	DU	DU	DU	DU	DU	DU
Option 2	CU	CU	DU	DU	DU	DU	DU	DU	DU
Option 3(2)	CU	CU	CU	DU	DU	DU	DU	DU	DU
Option 4(2)	CU	CU	CU	CU	DU	DU	DU	DU	DU
Option 5(3)	CU	CU	CU	CU	CU	DU	DU	DU	DU
Option 6(3)	CU	CU	CU	CU	CU	CU	DU	DU	DU
Option 7(3)	CU	CU	CU	CU	CU	CU	CU	DU	DU
Option 8(4)	CU	CU	CU	CU	CU	CU	CU	CU	DU

(1) It includes the HARQ.

(2) For this option, [10] does not specify the RRC allocation. It has been assumed to be in the CU.

(3) For this option, [10] does not list the functions performed by the CU but it just says: "Upper layer is in the central unit".

(4) In Option 8, the interface between PHY and RF is digital. A further analog PHY split option exists, known as RoF, not envisaged by [10]. It will be discussed in Chapter 7.

Options 3 is an intra-layer split option with two sublayers, Low RLC and High RLC. It has two variants: Option 3-1 and Option 3-2.

In Option 3-1, the low RLC sublayer performs segmentation functions and the high RLC sublayer performs ARQ and other RLC functions. In the AM operation (Section 3.3), all RLC functions are performed by the high RLC sublayer in the CU, except the segmentation which is performed by the low RLC sublayer in the DU. Having ARQ and packet ordering performed at the CU makes Option 3-1 robust to nonideal transport conditions. For example, recovering failures over the transport network may be possible using centralized ARQ mechanisms for protection of critical data and control signaling. Moreover, the centralization of the ARQ provides pooling gain and reduces the processing and buffering time in the DU. However, the ARQ centralization makes Option 3-1 latency sensitive, since the retransmission time is affected by the latency introduced by the transport network.

In Option 3-2, with reference to the DL, the low RLC sublayer includes TM, UM, and AM RLC transmitting functions, and the routing function of the AM receiver. The high RLC sublayer includes TM and UM RLC receiving functions, and the AM RLC receiving function, except for the routing function. In Option 3-2, the high RLC sublayer also performs the reception of the RLC status reports in UL. The configuration of Option 3-2 makes it less sensitive to the latency of the transport network between CU and DU.

Options 4 is just mentioned in [10], but no pros and cons are highlighted.

Option 5 splits the MAC into two sublayers: High MAC and low MAC. The high MAC sublayer performs centralized scheduling and control of multiple low MAC sublayers, and oversees the coordination among multiple cells to improve the interference management. Any time-critical function with stringent delay requirements is performed by the low MAC sublayer, including HARQ, processing of PHY measurements, and random access control. The bandwidth and latency impact of Option 5 on the transport network is moderate, since HARQ and cell-specific MAC functionalities are performed in the DU. Its main drawback is the complexity of the interface and the scheduling operation between CU and DU.

Option 6 moves the MAC completely into the CU. The interface between CU and DU carries data, measurements, and configuration-related information including layer mapping, beamforming, and resource block allocation. This option enables resource pooling and centralized scheduling for the MAC and the layers above. It introduces moderate traffic load on the transport network, since the payload only consists of TB bits, with no overhead from upper layers. A disadvantage of Option 6 is that it may require subframe-level timing interaction between the MAC in the CU and the PHY in the DU. Moreover, the round-trip delay in the transport network affects HARQ timing and scheduling.

Option 7 introduces a split point within the physical layer. Due to the complexity of the PHY processing chain (Figure 2.18), multiple implementations of this option are possible, including asymmetrical variants where the split point is different for UL and DL. Pros and cons of Option 7 are similar to Option 6, but better coordination features and pooling gain are gained in Option 7 at the expense of higher bandwidth and latency sensitivity. Some examples of Intra-PHY split are provided in [10]: In Option 7-1, the DU performs FFT and CP removal in UL, and IFFT and CP addition in DL, while all other PHY functions are performed in the CU; in Option 7-2, the DU performs FFT, CP removal, and resource demapping in UL, and IFFT, CP addition, resource mapping, and precoding in DL, while all other PHY functions are performed in the CU; in Option 7-3, applicable only to the DL, the encoder is in the CU, and the rest of PHY functions are in the DU.

Option 8 separates the RF layer from the PHY layer. It permits an extreme centralization and coordination of all mobile network processes, efficiently supporting features such as coordinated multipoint (CoMP), MIMO, load balancing, and mobility. However, it poses demanding latency and capacity requirements to the transport network, which may be difficult to satisfy with a cost-effective design.

3.7 The 5G Transport Network Architecture

3.7.1 RAN Logical Interfaces

The problem of mapping radio functional split options in the architecture of an optical transport network has been addressed by the ITU-T, [11]. Before presenting the ITU-T 5G transport architecture, it is useful to recall the logical interfaces defined by the 3GPP for the next-generation radio access network (NG-RAN), shown in Figure 3.7.

The NG-RAN architecture consists of a set of BSs (gNBs in the 5G terminology), which may be split between a CU and a DU. gNBs or CUs are connected to the 5G core network (5GC) via the NG interface. Two gNBs or two CUs can be connected through the Xn interface. CU and DU are connected via the F1 interface.

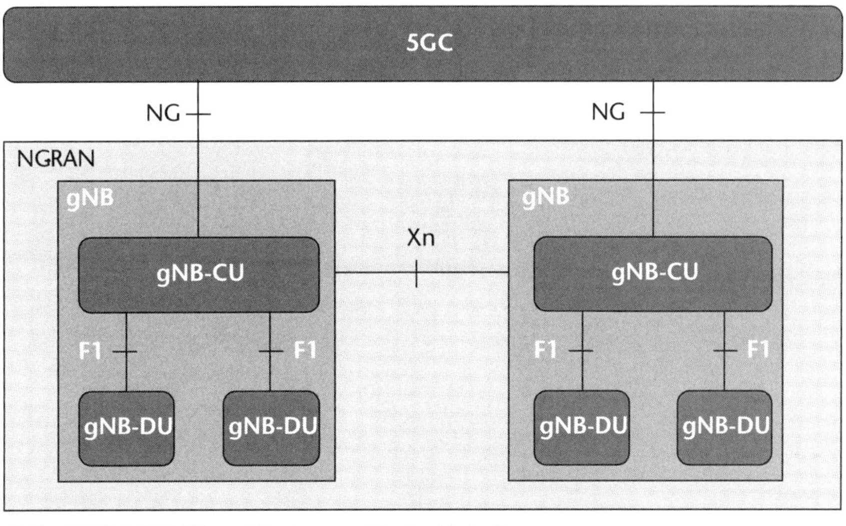

Figure 3.7 3GPP NG-RAN architecture and logical interfaces.

3.7.2 Definition of Fronthaul, Midhaul, and Backhaul

ITU-T adopted a transport architecture (Figure 3.8) consisting of three elements; CU, DU, and RU — sometimes indicated as remote radio unit (RRU) — whereas the 3GPP split architecture only considers CU and DU.

Fronthaul (FH) is defined as the transport network between RRU and DU; midhaul (MH) is defined as the transport network between DU and CU; and backhaul (BH) is the transport network between CU and core network (CN) or among CUs. The CN can be a 5GC or an LTE Evolved Packet Core (EPC) network. The backhaul links use the NG or Xn interface while the midhaul links use the F1 interface. There several possible fronthaul interfaces, which will be discussed in the following. Two examples are the CPRI [12] and eCPRI [13] interfaces defined by the Common Public Radio Interface (CPRI) industry cooperation.

It is necessary to highlight that the ITU-T terminology is not universally adopted, but a different nomenclature is used by different standardization organizations or is inherited from legacy systems. For example, the RRU is sometimes called remote radio head (RRH), and in LTE systems it just performs RF functions, according to Split Option 8. The equipment performing higher-level functionalities in LTE is usually indicated as baseband unit (BBU). To further complicate the picture, the BBU is called digital unit in some industry practice, known as the distributed unit in the ITU-T architecture. The CPRI cooperation introduced further terms, including radio equipment control (REC) and radio equipment (RE), visible as the left and right ends of the fronthaul link in Figure 3.8, respectively. Unless otherwise specified, the ITU-T definitions will be used in this book.

3.7.3 Mapping of Functional Split Options onto the Transport Network

In the mobile system generations prior to 5G, the term fronthaul referred to a split architecture based on Split Option 8, which separates the RF functionalities from all upper layers. The most used fronthaul interfaces were CPRI [12] and Open Base Station Architecture Initiative (OBSAI) [14]. Split Option 8 has the advantages of simplifying at most the design of the RU and allowing a deep centralization of the baseband processing functions, but it demands stringent latency and bandwidth requirements to the fronthaul link. Indeed, it is based on the transmission of

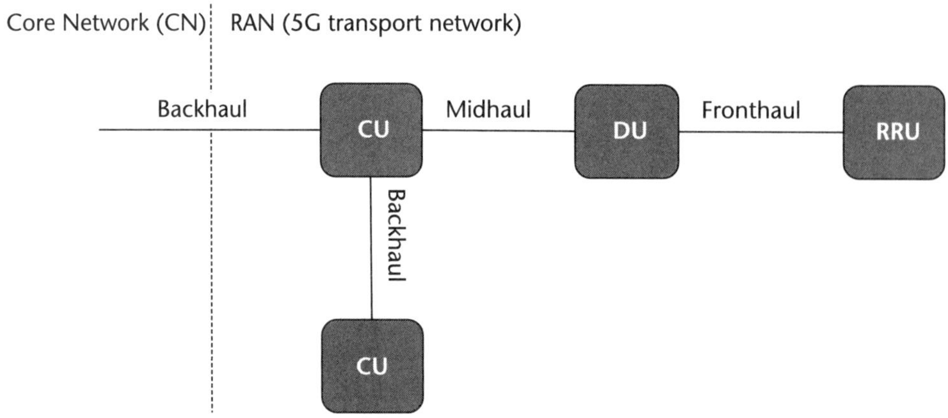

Figure 3.8 ITU-T definitions of fronthaul, midhaul, and backhaul.

time-domain digital samples of the two modulating signals at the I/Q branches of the RF modulator. At the receiver, the antenna signal is first sampled with sampling frequency given by (2.17) in Chapter 2, and then digitized with high resolution, typically 15 bits.

The calculation of the resulting bit rate will be made later in this chapter, but this technique had its disadvantages – it often leads to unreasonably high transport capacities in the presence of large bandwidths on air or massive MIMO systems. Values reported in [11] show the extent of the problem: Even without line coding, the bit rate of the fronthaul link is 2 Gbit/s for a 20 MHz LTE radio system with two antenna ports, increasing to 640 Gbit/s for a 200 MHz 5G 8x8 MIMO system with 64 antenna ports. Split option 8 has other disadvantages as well; for any option that incorporates the fronthaul link in the HARQ loop, the latency that can be allocated to the transport network is low, from tens to a few hundred of microseconds; additionally, Split Option 8 always requires continuous, or constant bit rate (CBR) transmission, even when there is no data to send. These considerations led to a deeper look at higher level split options for 5G fronthaul, keeping in mind that moving toward higher level splits will relax the latency and bandwidth requirements, but reduces the number of processing functions that can be centralized. It can never be emphasized enough that the right choice depends on the network scenario and the type of service. Nevertheless, limiting the number of options is necessary to avoid the market fragmentation and encourage standardized design practices, and to enable multivendor interoperability. Several standardization organizations moved in this direction. The 3GPP selected Split Option 2, between PDCP and high RLC, as the high-level split option. It is usually referred as F1 Interface, although F1 also indicates the generic interface defined by the ITU-T between CU and DU. No decision was taken by the 3GPP about a standardized low-level split point, for which there are two contenders, Option 6 (MAC–PHY split) and Option 7 (Intra-PHY split). According to [11], the low-level split interface will be generically indicated as Fx. The Small Cell Forum has extended the specification of its Functional Application Platform Interface (FAPI) with the addition of nFAPI [15]. nFAPI is a set of interfaces, based on virtualized Split Option 6, to allow multivendor interoperability and accelerate the deployment of small cells. Meanwhile, the CPRI consortium focused on the specification of intra-PHY split options (Split Option 7) based on data transport over packet networks, creating the eCPRI interface [13]. eCPRI introduces two possible split points in the DL, named I_D and II_D, and one in the UL, I_U, which roughly corresponds to Split Options 7-2 and 7-3. Among the other standardization bodies active on the specification of RAN interfaces, it is worth to mention the Metro Ethernet Forum (MEF) and IEEE P1914.1. MEF published its mobile backhaul implementation agreement [16] and is working on the definitions of next generation fronthaul. The IEEE P1914.1 standard [17] provides specifications on how to implement fronthaul over Ethernet networks.

Mapping examples of functional split options in the elements of a transport networks (CU, DU and RU) are shown in Figure 3.9, with the corresponding interfaces.

The first row in Figure 3.9 refers to LTE, where no MH interface is used. The CU and DU functions are performed by a single BBU and the RU is renamed RRH. The FH adopts Split Option 8, typically based on CPRI [12]. The other examples refer to 5G. In all, the BH interface to the 5GC is based on the NG interfaces and

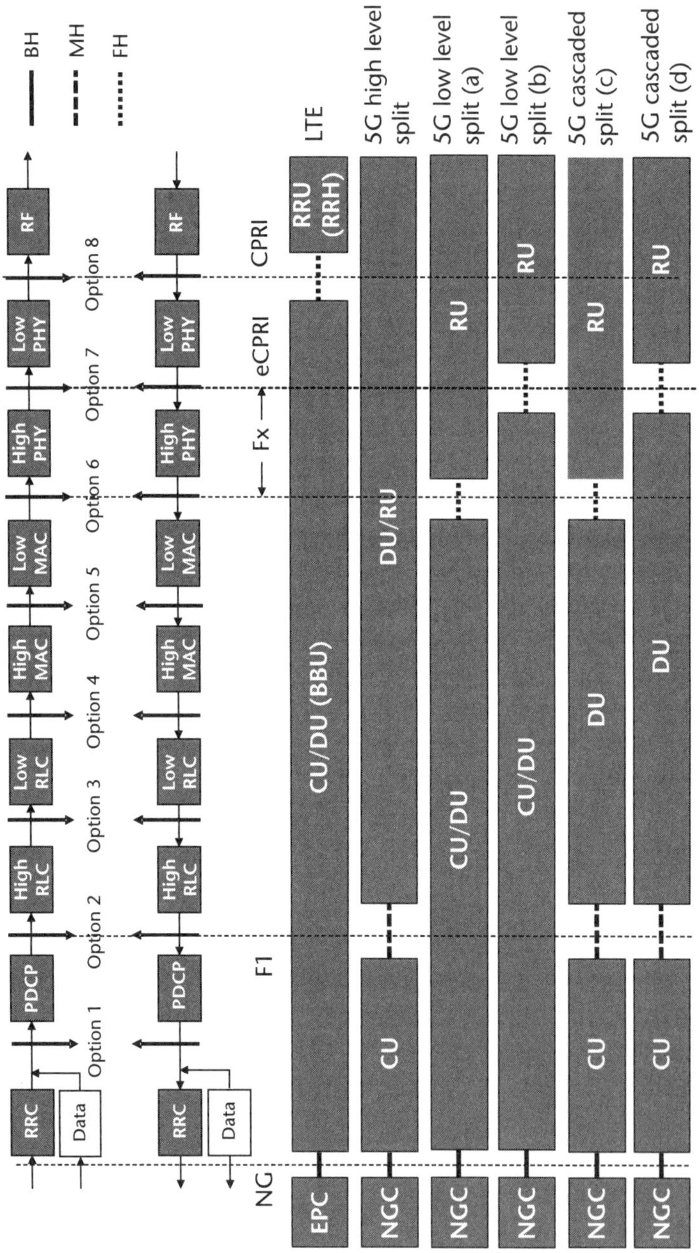

Figure 3.9 Example of mapping functional split options into the transport network.

the MH, when it is present, adopts the F1 interface as defined by the 3GPP. In the 5G high-level split scenario, no FH is present and DU and RU functions collapses in a single DU/RU unit. It could be the case of a centralized radio access network (C-RAN) where no strict coordination features are required, and no latency sensitive service is carried, but the centralization is driven by cost-saving considerations and the virtualization of high layers functions. Four low-level split examples are provided in Figure 3.9. In Figure 3.9(a), CU and DU functions are performed by a single CU/DU unit, and the FH is based on Split Option 6. It could be the case of a C-RAN where the latency requirements are more stringent than the previous case. Compared to it, the distance between CU/DU and RU will be shorter, and the geographical distribution of CU/DUs will be denser. The Figures 3.9(c) and (d) are cascaded split examples based on the ITU-T network architecture, which distributes the processing functions in three units. This allows a high deployment flexibility, centralizing higher level functionalities in a few CUs, and using more densely distributed DUs for low-level functions and latency critical services. In Figure 3.9(c), the split point is between MAC and PHY (moderate latency, higher centralization) whereas in the Figure 3.9(d) it is intra-PHY (low-latency, less centralization), such as based on eCPRI [13].

3.8 RAN Deployment Scenarios

As illustrated in the previous paragraph, the 5G transport network in Figure 3.1 can be segmented in fronthaul, midhaul and backhaul networks. There is not a single way to partition the network, but different deployment options are possible. Four macro scenarios [11] are described here:

1. Independent RU, CU, and DU locations [Figure 3.9(c) and (d)]. In this scenario, there are separate fronthaul, midhaul and backhaul networks; the maximum distance between RU and DU is typically 20 km while the distance between DU and CU is up to several tens of kilometers.
2. Co-located CU and DU [Figure 3.9(a) and (b)]. In this scenario, there is no MH, as in the LTE case (Figure 9, LTE example).
3. RU and DU integration (Figure 3.9, high-level split example). In this scenario, RU and DU are installed close to each other, maybe hundreds of meters, for example in the same building. To reduce cost, they are directly connected by a fiber cable and no transport equipment is needed. There is no FH link in this scenario.
4. RU, DU and CU integration. This is the case of a traditional BS where all the protocol stack layers (Figure 3.6) are monolithically integrated in the same equipment. Only BH is present in this scenario.

Figure 3.10 shows an example of transport network architecture with independent RsU, CU and DU locations (Scenario 1).

In Figure 3.10, the BH network is modeled as a multipoint-to-multipoint transport network based on interconnected rings (a mesh network topology would

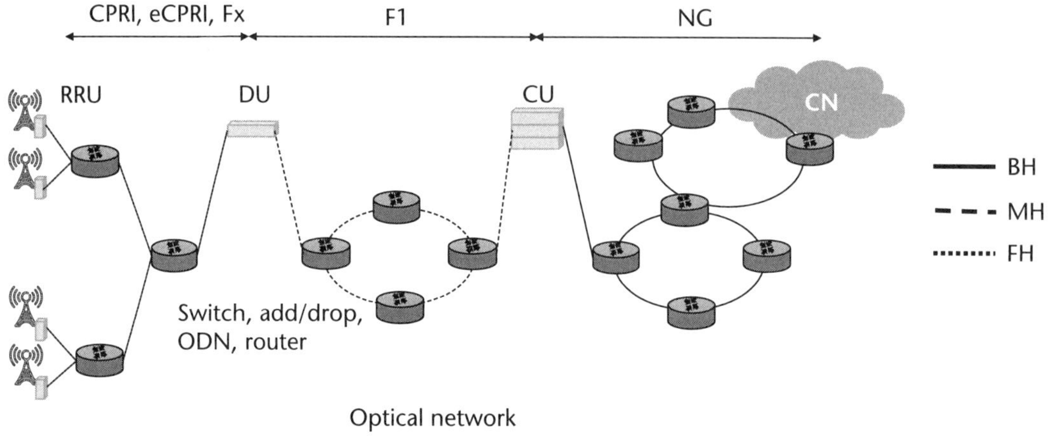

Figure 3.10 Transport network architecture with independent CU, DU, and RRU locations.

apply as well). In the FH and MH segments, the service is primarily point-to-point, that is, a DU only belongs to one CU (although occasional rerouting and reconfiguration is possible through switches, add/drop multiplexers, or routers) and an RRU only belongs to one DU. The FH network has a star or ring topology, either implemented through reconfigurable optical switches, or through wired optical distribution nodes (ODNs). Ring networks are expected to be typical instead in the MH segment, like in the current packet networks. A simplified scenario — where the BH and MH networks collapse in a single aggregation segment while the FH network remains confined in the access segment — is expected to be frequently adopted by operators. The main reason is that the typical latency requirements of MH and BH can be met by a regular packet switched network; this is not generally true for FH, where DU and RU are connected by low-latency fiber or wavelength point-to-point links, and statistical multiplexing has marginal benefits.

As typical distance reach values, [11] indicates 1–20 km for the FH segment and 20–40 km for the MH segment. Different ranges are provided for point-to-point BH (1–10 km); BH performed in an aggregation ring (5–80 km), for example, the first ring from the left (between CU and CN) in Figure 3.10; and second BH segment connected to the CN (20-300 km), such as the ring connected to the CN in Figure 3.10.

Reference [11] also provides coarse indications about the aggregate capacity. It spans from 25–800 Gbit/s for both MH and BH networks, whereas the FH capacity is calculated based on number of antenna ports and type of interface. Guidelines for the calculation of the FH capacity are provided in paragraph 3.10.

3.9 Network Slicing

In general, the transport network is a multiservice network, shared between 5G, legacy LTE, fixed access, among others. Not only it is necessary to provide isolation between each of these services, but it is also necessary to provide isolation between the different types of 5G services (see Table 3.1). This issue is often indicated as network slicing. A transport network is said to provide hard isolation when the traffic load in one slice has no impact on capacity and QoS of another slice. In packet networks, hard slicing is performed through the creation of virtual networks (VNs), where the forwarding plane ensures that the data of one VN are not accidentally delivered to a different VN and the throughput is properly dimensioned to prevent congestion situations that may lead to interference among VNs. One issue of packet based sliced networks is the necessity of over-dimensioning the switches to avoid the congestion. Giving up the advantages of the statistical multiplexing, hard isolation can be naturally implemented by circuit switched connections, either allocating dedicated wavelengths or through time division multiplexing.

3.10 Bit Rate and Latency with Different Functional Split Options

3.10.1 Bit Rate Dependency on the Split Option

It was noted in the previous paragraph that different functional splits correspond to different channel bit rates in the transport network. This paragraph provides guidelines to estimate such bit rate, with the warning that an exact calculation would require the knowledge of implementation details which may be vendor proprietary and not envisaged by the standard.

Three functional splits will be considered [18]: Split Option 8, Split Option 7-2 and Split Option 6. In the eCPRI specification [13] these options are indicated with the letters E, II_D/I_U, and D, respectively (Figure 3.11). In the following discussion, it will be referred to the DL direction of Figure 3.11.

Split Option 8 requires the highest bit rate, since after the IFFT, digitized time-domain samples are transmitted with a high resolution in number of bits per sample, due to the high PAPR of the OFDM signal. Moreover, the transmission of the digitized samples occurs even in absence of information data, uselessly occupying bandwidth. Both issues are solved with the two higher split options, where the transmitted signal is proportional to the OFDM modulation symbols, which require less resolution to be digitized, having finite discrete levels, and are present only if there is information data to transmit.

The bit rate difference of the three considered options increases in MIMO systems. In Split Option 8, the individual radiating elements of the antenna array are fed with independent signals, each one carrying digitized time-domain samples. In Split Option 7-2, the beamforming is performed at the RU (Figure 3.11) so that a single signal at the interface between DU and RU is sufficient for all the radiating elements that contribute to form the beam; then, phase shifted copies of that signal are generated inside the RU and sent to each radiating element. Further bit rate reduction occurs with Split Option 6, where the layer mapping function is moved into the RU, so that the signal is proportional to the number of layers (Section 2.4) rather than to the number of antenna elements. For example, an 8×8 antenna array could be divided in 8 subarrays, each subarray corresponding to a layer and consisting of the 8 radiating elements of a column [19]. With Split Option 8, the bit rate would be proportional to the number of radiating elements, or 64, while with Split Option 6, it is proportional to the number of layers, or 8.

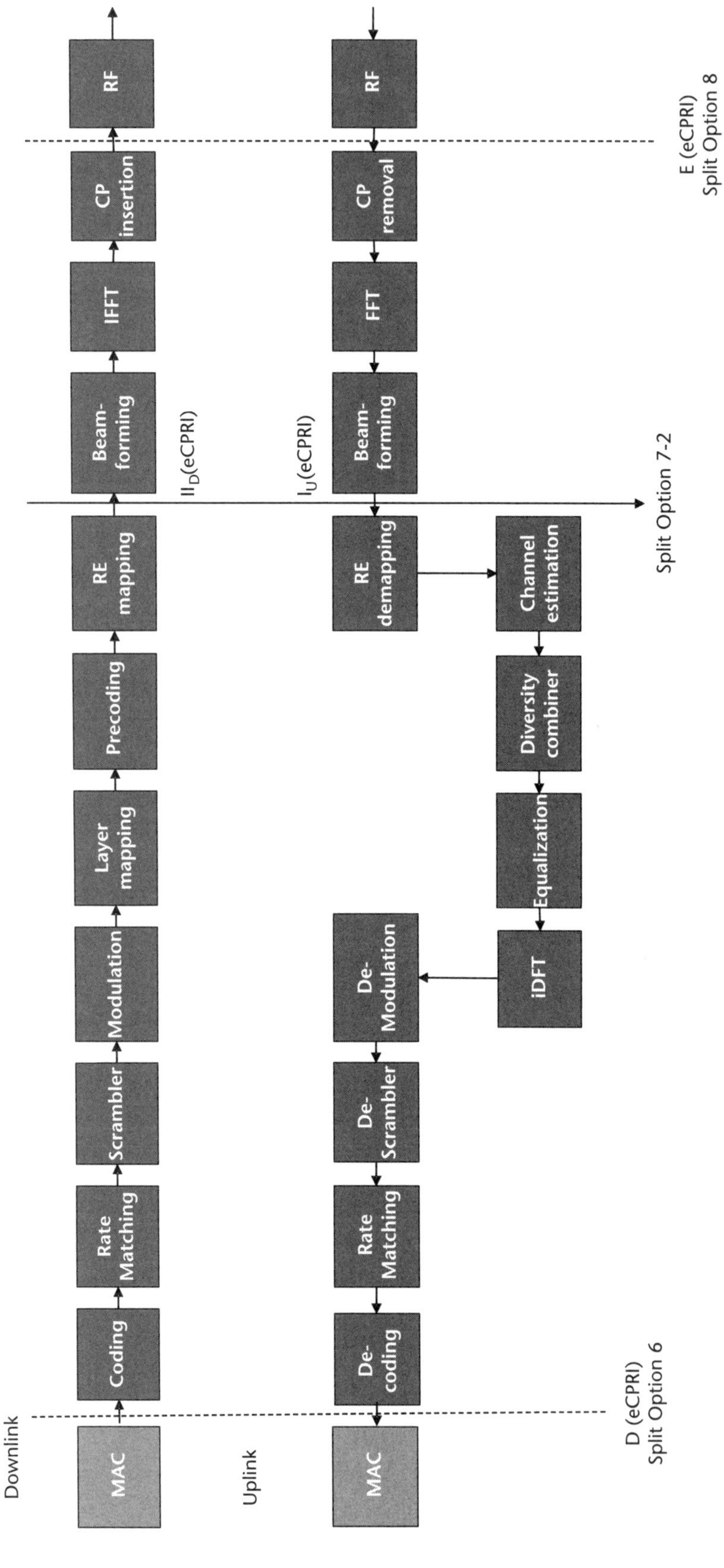

Figure 3.11 Split options in the eCPRI specificaton.

3.10.2 Bit Rate Calculation

In Split Option 8, the signal at the antenna port is sampled and quantized, independent of the presence of modulation symbols carrying information data. Hence, the bit rate R_b is given by:

$$R_b = 2 \cdot A \cdot f_s \cdot N_b \cdot OH_{cb} \cdot OH_{lc} \tag{3.2}$$

where:

The factor 2 takes into account the separate processing of I/Q samples;

A is the number of antenna ports;

f_s is the sampling frequency;

N_b is the resolution, in number of bits, of the time-domain quantizer;

OH_{cb} is the channel overhead of the transport interface;

OH_{lc} is the line coding overhead of the transport interface.

For a 20 MHZ LTE signal carried over CPRI interface [12], it is A=2, f_s=30.72 MHz (Table 2.2), N_b=15, OH_{CH}=16/15 and OH_{LC}=10/8 (using an 8B10B line code). The resulting bit rate is 2.4576 Gbit/s, which exactly corresponds to line bit rate option 3 in [12]. In [12], the bit rate is calculated in a slightly different way, as 2 (IQ samples) × 2 (antenna ports) × 491.52 MHz × 10/8 (line code overhead). The frequency value of 491.52 MHz is an exact multiple of both LTE sampling frequency (16 × 30.72 MHz) and UMTS chip rate (128 × 3.84 Mbit/s). This explains the 16/15 factor in the channel overhead OH_{CH} of (3.2), which is used to match this frequency value with a quantizer resolution of 15 bits.

The necessity to consider alternative split options in 5G can be understood applying (3.2) to a MIMO system with μ=4, f_s=491.52 MHz (Table 2.5), A=8 (may be the case of 8 layers), OH_{CH} =1.07 (this is the FEC overhead of a standard OTN channel [1]) and OH_{LC}=66/64 (using a 64B66B line code). The resulting bit rate is 130.17 Gbit/s, which requires expensive optical transmission techniques, such as dual polarization transmission with coherent detection, to be carried over a transport network.

In Split Option 7-2, the FTT function is moved into the RU so that the bit rate is proportional to the modulation symbol rate and it becomes zero if there is no data to transmit [see (3.3)]:

$$R_b = \frac{2 \cdot A \cdot N_{actual_{SC}} \cdot N_b \cdot OH_{cb} \cdot OH_{lc}}{T} \cdot r \tag{3.3}$$

where:

The factor 2 takes into account the separate processing of I and Q samples;

A is the number of antenna ports (in absence of beamforming, see section 3.10.1);

$N_{actual\ SC}$ is number of occupied subcarriers (Table 2.2);

N_b is the resolution, in number of bits, of the frequency domain quantizer;

OH_{cb} is the channel overhead of the transport interface;

OH_{lc} is the line coding overhead of the transport interface;

T is the symbol time;

r is the average traffic load, defined as the ratio between the time interval data are transmitted within and the total transmission time (r=1 if data are always transmitted).

In the MIMO system example already considered for Split Option 8, it is N_{actual_SC}=138 and T=4.17μs (Table 2.5). Assuming N_b=9 for the quantizer in the frequency domain, the bit rate is 63.1 Gbit/s, less than a half the bit rate obtained with Split Option 8. A second example is an 8-layer MIMO system consisting of five frequency division multiplexed LTE signals. For such signal T = 66.67 μs and N_{actual_SC}= 5 × 1,200 =6,000. Keeping the other parameters as in the previous example, R_b = 14.3 Gbit/s. With a 66B64B-coded split option 8 channel, the same signal would require 40.5 Gbit/s (2 × 30.72 MHz × 15 × 16/15 × 66/64 × 5 FDM channels × 8 layers), which could be accommodated by time division multiplexing two 24.33 Gbit/s CPRI channels, or four 10.14 Gbit/s CPRI channels. In both examples, further overhead should be allocated for control signaling and for sending the beamforming weights. Moreover, an exact analysis should also consider the frame structure of the various type of radio channels, but these are implementation details outside of the scope of this book.

Moving the split point to the edge between PHY and MAC (Split Option 6), the difference between transport channel bit rate and bit rate on air shrinks further, [as in (3.4)]:

$$R_b = \frac{A \cdot N_{actual_{SC}} \cdot R_{code} \cdot log_2(M) \cdot OH_{cb} \cdot OH_{lc}}{T} \cdot r \tag{3.4}$$

where:

A is the number of layers;

$N_{actual\ SC}$ is number of occupied subcarriers;

R_{code} is code rate;

M is the number of modulation symbols in the constellation;

OH_{cb} is the channel overhead, according to the adopted transport interface;

OH_{lc} is the line coding overhead, according to the adopted transport interface;

T is the symbol time;

r is the average utilization of the subcarriers (traffic load).

Assuming, for the MIMO system considered above, R_{code}=0.85 and 256-QAM modulation, the bit rate is 23.8 Gbit/s.

To facilitate the capacity dimensioning of the transport network and make it independent of nonpublicly available vendor proprietary assumptions, [11] collected the inputs of industries and operators to provide indicative bit rate values of the various transport interfaces. They are summarized in Table 3.4.

3.10.3 Latency Calculation

The latency budget of a radio system was analyzed in Section 3.5. In Figure 3.5, the total time available for the transmission between UE and eNodeB is 1 ms, but different values are possible, depending on the type of service. For example,[11] reports 4 ms for eMBB and 0.5 ms for URLLC.

In a split architecture, the transmission time includes not only the propagation delay on air but also the delay introduced by the transport network between DU and CU, primarily due to the propagation in optical fiber and the queuing times in the packet switches. To understand how the transport latency budget is affected by the functional split between CU and DU, it is convenient to recall Table 3.2, whose values are reported in Table 3.5, with the following modifications: All times are given in TTI units to make the discussion more general; the processing of transmission requests is assumed to be handled by the CU; only the time interval from an UE request to the start of transmission of UL data is considered.

The time diagram in Figure 3.12 shows request flow and timelines between the various units; T_L is the latency of the transport network between CU and DU.

Assuming that the HARQ handles N processes in parallel so that N TTI intervals can be allocated for the retransmission, the following equation holds:

$$TTI + T_L + T_{CU} + T_L + TTI + T_{UE} = N \cdot TTI \tag{3.5}$$

Hence, the round-trip latency of the transport network, RTT_T, is given by:

Table 3.4 Channel Capacity for Different Interfaces

F1 Interface (Split Option 2)	4.016 Gbit/s (DL), 3.024 Gbit/s (UL)
Fx Interface (Split Option 7-1)	10.1–22.2 Gbit/s (DL), 16.6–21.6 Gbit/s (UL)
Fx Interface (Split Option 7-2)	37.8–86.1 Gbit/s (DL), 53.8–86.1 Gbit/s (UL)
Xn Interface	25–50 Gbit/s
NG Interface	10–25 Gbit/s for CU, >100 Gbit/s for CN
FH (Split Option 8)	Use (3.3) From ~2 Gbit/s (2 antenna ports, 20 MHz BW) to ~3.2 Tbit/s (64 antenna ports, 1 GHz BW)

Table 3.5 Latency Contributions

Description	*Time*
UE sends a scheduling request	TTI
CU decodes the scheduling request and generates a scheduling grant, T_{CU}	3·TTI
Transmission of the scheduling grant	TTI
UE processing delay (decoding of scheduling grant + PHY encoding of UL data), T_{UE}	3·TTI
Transmission of UL data	TTI

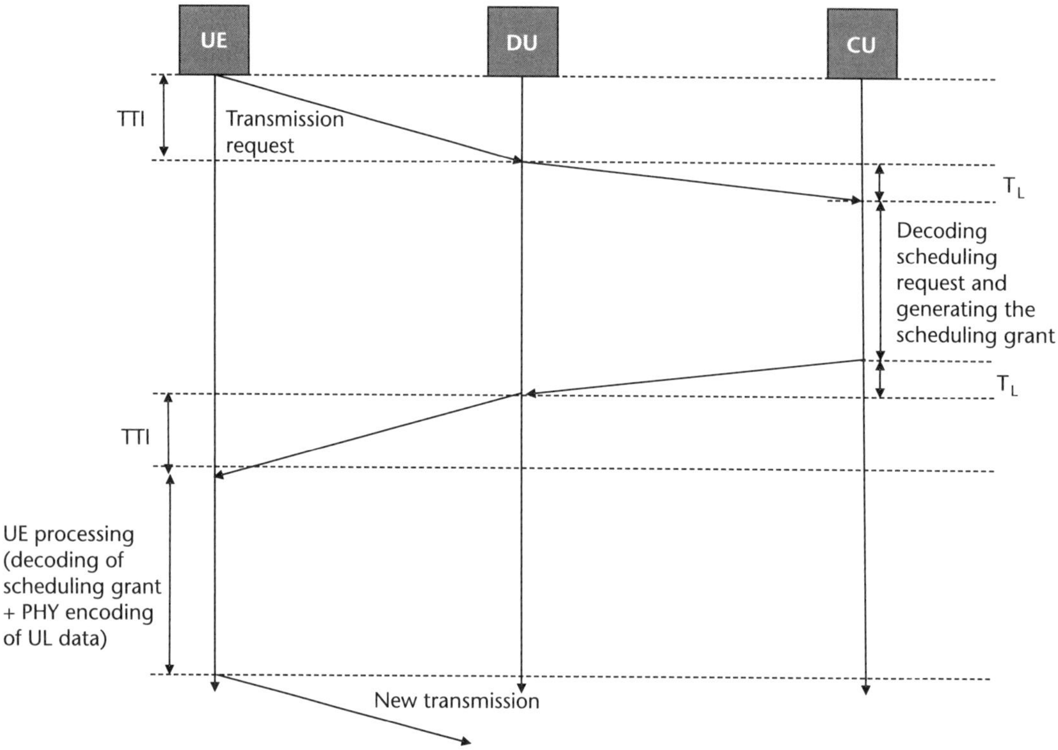

Figure 3.12 Time diagram of a transmission request in a split architecture.

$$RTT_T = 2 \cdot T_L = (N-2) \cdot TTI + T_{CU} + T_{UE} \tag{3.6}$$

Replacing the values of Table 5 in (3.6), it is obtained:

$$RTT_T = (N-8) \cdot TTI \tag{3.7}$$

Equation (3.7) shows that a shorter TTI reduces the latency budget, as expected. It also gives a lower bound for the number of HARQ processes that can managed in parallel, which is 8 in this example [20].

There is no exact rule for partitioning the end-to-end transmission time between a radio communication system and a transport network connected to it. Allocating more latency to the transport network imposes stricter requirements on the processing times of CU and DU. On the other hand, reducing the transport latency budget decreases the fiber distance between CU and DU and requires special scheduling rules to avoid long queues in the switches. The adopted functional split plays a major role in the choice: Split options where the HARQ is managed by both CU and DU have tight latency requirements, since the delay introduced by the transport network is part of the retransmission time. For the reasons above, reaching an agreement on the latency values is not easy, especially considering that the wireless network and the optical network often point to different departments in the operators' and system vendors' organizations. There is a general consensus on setting the maximum one-way latency value introduced by an LTE fronthaul network based on CPRI to 100 μs. Since the propagation delay introduced by the optical fiber is 5 μs/km, this corresponds to a maximum distance of 20 km between CU and DU. eCPRI [21] defined four different latency classes, including both fiber and switching delay (Table 3.6).

More relaxed values of latency are allowed with higher level split options. For example, [11] reports 1.5–10 ms as latency range for the F1 interface.

In a real transport network, the latency is known with some degree of approximation, called delay estimation accuracy, or time error (TE). The TE must be lower than the time alignment error (TAE), defined as the relative time error between an arbitrary pair of transmitting antenna ports in a mobile network. The TAE specified in LTE equals two sampling times, 65 ns:

$$TAE = \frac{2}{f_s} \tag{3.8}$$

eCPRI defines four categories of TEs. Each category has subcases related to the synchronization mechanisms used in the network. Without going into such level of detail, Table 3.7 shows the minimum TE values specified by eCPRI.

The values in Table 3.7 are defined at the interface between the transport network and a DU or a CU, category A+, A, and B values are relative time values, defined between a pair of interfaces. Category C values are instead defined with respect to an absolute timing reference.

Any delay difference between the UL and the DL is a source of TE. In a transport network, such asymmetric delay can have several causes: Length difference of the UL and DL optical fibers (65 ns of delay approximately correspond to 13 meters of fiber); use of different wavelengths in UL and DL, which undergo a different propagation delay due to the fiber chromatic dispersion; asymmetries introduced by framers, multiplexers or FEC chipsets in optical transceivers; and queuing times

Table 3.6 eCPRI Latency Classes

Latency Class	*Latency*	*Use Xase*
High25	25 μs	Ultralow latency performance
High100	100 μs	For full LTE or NR performance
High200	200 μs	For installations where the lengths of fiber links are in the 40 km range
High500	500 μs	Large latency installations

Table 3.7 Minimum Time Error Specified by eCPRI

Category	*Minimum TE*	*TAE*
A+	20 ns	65 ns
A	60 ns	130 ns
B	100 ns	260 ns
C	1100 ns	3 μs

in packet switches. The contribution due to the length difference can be easily calculated from the optical fiber delay, which is 5 μs/km. The delay asymmetry due to the chromatic dispersion can be estimated using (3.9) from [22] which gives the chromatic dispersion coefficient D in ps/(nm·km), versus the wavelength λ in nm:

$$D(\lambda) = \frac{S_0}{4}\left(\lambda - \frac{\lambda_0^4}{\lambda^3}\right) \tag{3.9}$$

In (3.9), λ_0 is the zero-dispersion wavelength, whose minimum value is 1,300 nm for the single mode standard fiber specified in [22], and S_0 is the dispersion slope at λ_0, having the maximum value of 0.092 ps/(nm^2·km). The fiber delay τ, in ps/km, versus the wavelength, is obtained integrating (3.9):

$$\tau(\lambda) = \tau_C + \frac{S_0}{8}\left(\lambda^2 + \frac{\lambda_0^4}{\lambda^2}\right) \tag{3.10}$$

The unknown constant τ_C cancels out when calculating the delay difference between two UL and DL optical channels that use different wavelengths, λ_{UL} and λ_{DL}, respectively:

$$\Delta\tau_{UL-DL} = \left|\tau(\lambda_{UL}) - \tau(\lambda_{DL})\right| \tag{3.11}$$

In CPRI-based FH networks, $\Delta\tau_{UL-DL}$ could lead to an undesired delay imbalance between time synchronized RRUs connected to the same CU. As maximum tolerable value of $\Delta\tau_{UL-DL}$, it is often assumed ±8.138 ns, which is actually the value that [12] specifies for the link delay accuracy in DL between master and slave service access points, excluding the cable length.

3.11 Summary

A long journey began in this chapter. Starting from the protocol stack of a mobile access network, different options of split architecture were analyzed. Then, corresponding network elements and interfaces were defined for a transport network connected to the mobile network. The bit rate values for the principal transport interfaces were calculated based on the air interface specification discussed in Chapter 2. Then, the latency budget was analyzed after having introduced the retransmission mechanisms in the mobile network, and the timing error sources introduced by the transport network. In the next chapter, the design issues of an optical transport network able to comply with all the requirements analyzed in this chapter will be addressed.

References

[1] Interfaces for the Optical Transport Network, Recommendation ITU-T. G.709/Y.1331.

[2] Osseiran, A., J. F. Monserrat, P. Marsch, et al., *5G Mobile and Wireless Communications Technology*, MA: Cambridge University Press, 2016.

[3] "5G Systems - Enabling the Transformation of Industry and Society," Ericsson White Paper, January 2017.

[4] Dahlman, E., S. Parkvall, J. Sköld, et al., *3G Evolution - HSPA and LTE for Mobile Broadband*, MA: Academic Press, 2008, pp. 119–123.

[5] Semenov, S., "Modified Maximal Ratio Combining HARQ Scheme for HSDPA," *Proc. 2004 IEEE 15th International Symposium on Personal, Indoor and Mobile Radio Communications*, January 2005.

[6] Bertenyi, B., S. Nagata, H. Kooropaty, et al., "5G NR Radio Interface," *Journal of ICT*, Combined Special Issue 1 & 2, River Publishers, 2018.

[7] "Feasibility Study for Further Advancements for E-UTRA (LTE-Advanced)," 3GPP Technical Specification 36.912.

[8] Takeda, K., L. H. Wang, and S. Nagata, "Latency Reduction toward 5G IEEE Wireless Communications," *Wireless Communications,* Vol. 24, No. 3, June 2017.

[9] Teyeb, O., G. Wikström, M. Stattin, et al., "Evolving LTE to Fit the 5G Future," *Ericsson Technology Review,* 2017.

[10] "Study on New Radio Access Technology: Radio Access Architecture and Interfaces," 3GPP Technical Specification 38.801.

[11] "Transport Network Support of IMT-2020/5G," Technical Report ITU-T GSTR-TN5G.

[12] "CPRI Interface Specification", http://www.cpri.info/spec.html.

[13] "eCPRI Interface Specification", http://www.cpri.info/spec.html.

[14] "BTS System Reference Document v2.0", OBSAI Specification, www.obsai.com.

[15] "nFAPI and FAPI Specifications," Doc. Number SCF082, Small Cell Forum, May 2017.

[16] Mobile Backhaul Phase 3, MEF Implementation Agreement 22.2, January 2016.

[17] IEEE1914.1, "Standard for Packet-Based Fronthaul Transport Networks," http://sites.ieee.org/sagroups-1914/.

[18] Bartlet, J., N. Vucic, D. Camps-Mur, et al., "5G Transport Network Requirements for the Next Generation Fronthaul Interface," *EURAISIP Journal on Wireless Communications and Networking,* May 2017.

[19] Von Butovitsch, P., D. Astely, A. Furuskär, et al., "Advanced Antenna Systems for 5G Networks," Ericsson White Paper, November 2018.

[20] Cai, L., and A. Tazi, "5G Transport Latency Requirement Analysis," *Proc. IEEE 1914 NGFI Working Group meeting,* Dallas, TX, April 2017.

[21] eCPRI Transport Network, Requirements Specifications, http://www.cpri.info/spec.html.

[22] Characteristics of a Single-Mode Optical Fibre and Cable, Recommendation ITU-T. G.652.

CHAPTER 4

Optical Transmission Modeling in Digital RANs

4.1 Introduction

The previous chapter analyzed architecture and requirements of a digital RAN. This chapter will illustrate how to model the optical links in the network and what the main design parameters are. Performance metrics, transmitter and receiver models, and propagation in fiber will be discussed in the simplest case of point-to-point transmission between two ports of two different units in a RAN, for example, a RU and a DU connected by a dedicated fiber link. Dedicated fiber links are used in the deployment scenario referred as distributed RAN (D-RAN). The models will be extended to more complicated scenarios in the next chapter.

The propagation over a point-to-point fiber link can be unidirectional, with two different fibers allocated for the DL and the UL [Figure 4.1(a)] or bidirectional, on a single fiber shared by UL and DL signals transmitted over separate optical carriers [Figure 4.1(b)].

Bidirectional links use duplexers to separate the DL signal from the UL signal. A duplexer is a 3-port optical filter, whose working principle is illustrated in Figure 4.2. It is typically embedded in the transceiver and has a typical insertion loss of about 1 dB. Its ports are all bidirectional so that the same device applies to both the ends of the link.

The bands used in fiber communication systems are defined in [1] and reported in Table 4.1.

In optical communications, frequency and wavelength units are used in interchangeable way. Equation (4.1) is used to convert the value of an optical carrier from meter to hertz:

$$f = \frac{c}{\lambda} \tag{4.1}$$

In (4.1), c is the light speed *in vacuum*, 299,792,458 m/s, even if the actual light speed in fiber is c/n, where n is the effective refractive index of the main

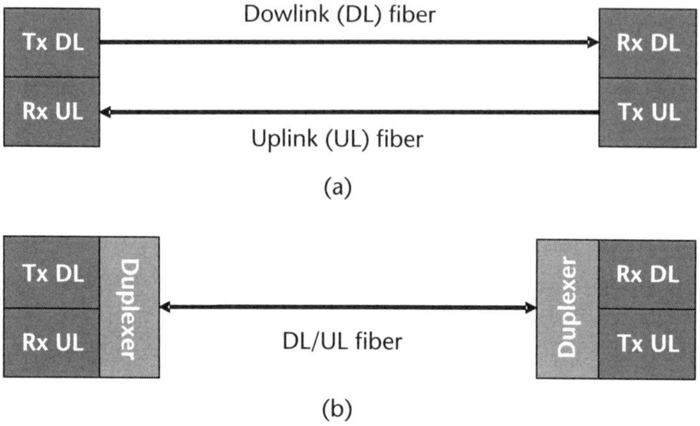

Figure 4.1 (a) Unidirectional and (b) bidirectional point-to-point fiber link.

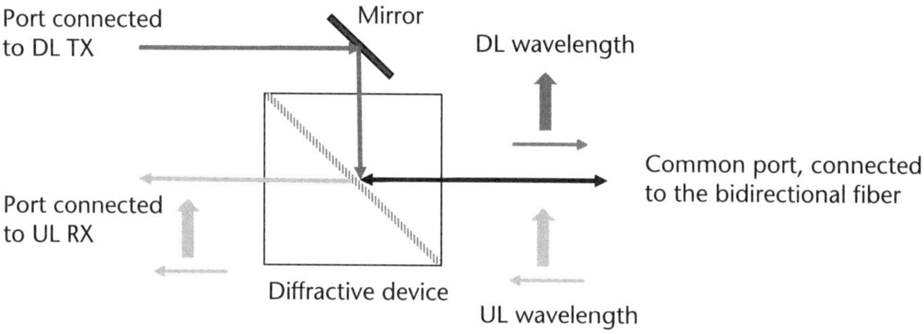

Figure 4.2 Duplexer working principle.

Table 4.1 Fiber Communications System Bands

Band	*Name*	*Range (nm)*	*Range (THz)*
O band	Original	1260 to 1360	220.43 to 237.93
E band	Extended	1360 to 1460	205.34 to 220.43
S band	Short wavelength	1460 to 1530	195.64 to 205.34
C band	Conventional	1530 to 1565	191.56 to 195.94
L band	Long wavelength	1565 to 1625	184.49 to 191.56
U band	Ultra-long wavelength	1625 to 1675	178.98 to 184.49

propagation mode in fiber, which is about 1.5. To convert values of optical bandwidth from meter to hertz, intended as a small wavelength interval around a much larger optical carrier, it is necessary to differentiate (4.1):

$$\Delta f = -\frac{c}{\lambda^2}\Delta\lambda \tag{4.2}$$

Applying (4.2) to the C band, which is centered around 1,547.5 nm, it can be verified that a bandwidth of 0.1 nm corresponds to 12.5 GHz. Thus, 100 GHz and 50 GHz, which are the frequency spacing values defined in [2] for the adjacent channels of a DWDM system, correspond to 0.8 and 0.4 nm respectively.

4.2 Fiber Attenuation

The optical fiber attenuates the signal power transmitted into an optical link. The attenuation coefficient, , is measured in dB/km, and varies with the wavelength, as shown in Figure 4.3 [3]. In many equations of this chapter, α is expressed in linear units, or km^{-1}. For conversion from decibels to linear units, and vice versa, (4.3) is used.

$$\alpha_{linear\ units} = \frac{\alpha_{dB}}{10 \cdot \log_{10}(e)} \sim \frac{\alpha_{dB}}{4.343} \tag{4.3}$$

The curve in Figure 4.3 is qualitative: Different fibers, or even different batches of fibers produced by the same manufacturer, have a different attenuation coefficient. For example, due to improvements of the production process, new fibers do not present the water absorption peak, visible in Figure 4.3 around 1,383 nm.

Figure 4.3 also shows the wavelength ranges, or transmission windows, used for optical fiber transmission. Optical transmitters were initially developed in the first window, which is still used for short distance communications, due to the availability of low-cost directly modulated optical sources. However, in the first window, both multimode propagation and high attenuation coefficients severely limit the achievable link distance. Higher distances are achievable in the second window, where both the dispersion effects and attenuation coefficient are lower. The fiber cut-off wavelength, that is, the wavelength at which the propagation switches from multimode to single mode, is in the second window: A typical value is 1,260 nm. The third window is used in DWDM systems, due to the low attenuation coefficient

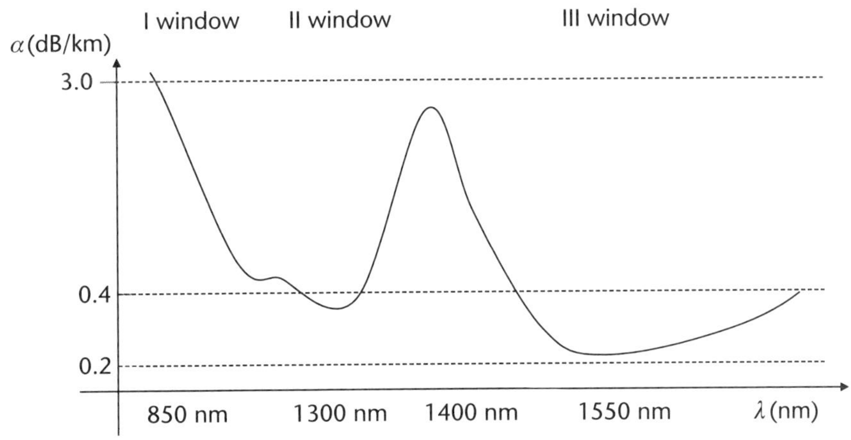

Figure 4.3 Fiber attenuation coefficient

and the availability of erbium-doped fiber amplifiers (EDFAs) that operate in this wavelength range. In third window, the chromatic dispersion is high, but it can be compensated by placing in the link a special kind of optical fiber, called a dispersion compensating fiber (DCF), which has the sign of the chromatic dispersion coefficient opposite to the sign of the fiber dispersion coefficient.

In an optical link, additional attenuation is introduced by fiber splices and connectors, present since the installation or added following a cable repair. Moreover, the exact attenuation coefficient of fibers installed long time ago may be unknown even to the infrastructure owner. To face all these uncertainties, ITU-T defined worst case values of attenuation coefficient, equal to 0.275 dB/km in C band and 0.55 dB/km in O band [1]. These values include splices, connectors, optical attenuators and additional cable margin to cover future additional splices, increased cable length, variations due to environmental factors, degradation of connectors, and other values. If the number of splices and connectors is known, the attenuation of the optical link can be calculated as:

$$A = \alpha L + \alpha_S x + \alpha_C y \tag{4.4}$$

where:

A is the optical link attenuation, in decibels;

α is the fiber attenuation coefficient, in dB/km;

α_S is the mean splice loss;

x is number of splices in the link;

α_C is mean loss of connectors;

y is number of connectors in the link;

L is the link length.

The attenuation budget of a point-to-point optical link is:

$$P_{RX-min} = P_{Tx-min} - A - P_{path} \tag{4.5}$$

In (4.5), $P_{Tx\text{-}min}$ is the minimum value of transmitted optical power, averaged over a long sequence of modulation symbols. It is guaranteed by the transmitter supplier, up to the system end of life (EOL), and over the full operating temperature range, for example –40°–85°C. $P_{Rx\text{-}min}$ is the receiver sensitivity, defined as the minimum value of received optical power at which a given BER is guaranteed. Typically, it is assumed a BER of 10^{-12} in absence of FEC. In presence of FEC, it is taken as reference for the BER value, before the error correction (pre-FEC BER), that guarantees a BER lower than 10^{-12} after the error correction (post-FEC BER). For a 10 Gbit/s OTN channel [4], the pre-FEC BER is slightly higher than 10^{-5}. The sensitivity is guaranteed by the receiver vendor up to the system EOL, over the full operating temperature range, and with worst-case transmitter and link parameters. P_{Path} is the maximum optical path penalty. It is defined as the apparent reduction of receiver sensitivity due to the distortion of the signal waveform during its transmission over the link. The path penalty is allocated by the designer to include any power penalty associated with the optical path, such as chromatic dispersion, polarization mode dispersion, reflections, etc. Typical path penalty values are in the range from 1–2 dB.

4.3 Performance Metrics in Optical Communication Systems

Several performance metrics are used in optical communications. The most used ones are the BER, the Q factor, the optical modulation amplitude (OMA), the error vector magnitude (EVM) and the optical signal-to-noise ratio (OSNR).

4.3.1 Bit Error Rate

The BER is the ultimate performance indicator for any digital communication system. It is defined as:

$$BER = \lim_{N \to \infty} \frac{n_E}{N} \tag{4.6}$$

Where n_E is the number or errored received bits and N is the number of transmitted bits. Since the measurement time is finite, the BER can only be estimated with a certain confidence level, C, given by (4.7):

$$n = \frac{ln(1-C)}{\ln(1-BER)} \tag{4.7}$$

In (4.7), n is the number of error-free bits required to achieve a desired confidence level, C. For example, at a BER equal to 10^{-12}, it is necessary to detect $3{\cdot}10^{12}$ error-free bits to achieve a 95% confidence level. The corresponding measurement time is 20 mins at a bit rate of 2.5 Gbit/s and 5 mins at 10 Gbit/s. Equation (4.7) allows the BER accuracy to be estimated at low bit rates, at which the number of transmitted bits over the measurement time is low, and at low BER, which requires a high number of errored bits to be estimated. At low bit rates, (4.7) suggests using a sequence of error-free bits to decide with a certain confidence level that the BER is below a given threshold.

4.3.2 Q Factor

The BER of a digital communication system results from all propagation impairments and noise sources that affect the signal from transmission to reception, but just measuring the BER, it is impossible to distinguish the individual contributions of each effect. Alternative metrics, like the Q factor, are more suitable at this purpose. For a binary intensity-modulated signal, the Q factor is defined as:

$$Q = \frac{\mu_1 - \mu_0}{\sigma_1 + \sigma_0} \tag{4.8}$$

where μ_1 and μ_0 are the mean values of the received signal, when marks and spaces are transmitted, respectively, and σ_1 and σ_0 are the its standard deviations. Often, the Q factor is expressed in decibels:

$$Q_{dB} = 20 \cdot log_{10}(Q) \tag{4.9}$$

The definition of Q factor can be generalized to multilevel modulation formats, like Order-4 Pulse Amplitude Modulation (PAM-4), defining a different Q factor value for each possible pair of signal levels. For a binary amplitude modulated signal affected by additive gaussian noise and receiver set at the optimal decision threshold, there is a direct relation between BER and Q factor:

$$BER = \frac{1}{2} erfc\left(\frac{Q}{\sqrt{2}}\right) = \frac{1}{2\sqrt{2\pi}} \int_x^{\infty} e^{-\frac{\xi^2}{2}} d\xi \tag{4.10}$$

For example, it is Q =7.03, or 16.9 dB, at BER=10^{-12}. It may be useful to use closed-form approximations of (4.10) [5]. Two examples, valid for Q>3, are reported below:

$$BER \sim \frac{e^{-\frac{Q^2}{2}}}{\sqrt{2\pi}Q} \qquad BER \sim \frac{e^{-\frac{Q^2}{2}}}{\sqrt{2\pi}\left[\left(1 - \frac{1}{\pi}\right)Q + \frac{\sqrt{Q^2 + 2\pi}}{\pi}\right]} \tag{4.11}$$

In a binary amplitude modulated communication system, the optimum value of the decision threshold is:

$$I_{th-opt} = \frac{\mu_1\sigma_0 + \mu_0\sigma_1}{\sigma_0 + \sigma_1} \tag{4.12}$$

Both (4.10) and (4.12) assume additive gaussian noise, a hypothesis that is no longer valid in optically preamplified systems [6]. For those systems, it can be demonstrated that (4.10) still holds at the optimal decision threshold, which is however different from the value given by (4.12) [7]. A practical method to optimize the decision threshold of a receiver is moving it until balancing the number of errors corresponding to transmitted 0 and 1 bits. This can be done by using the error count function provided by the FEC decoder. However, in some receivers the decision threshold is not adjustable. A typical case is the alternate current (AC)-coupled receiver in Figure 4.4, where the capacitor after the photodiode sets the average

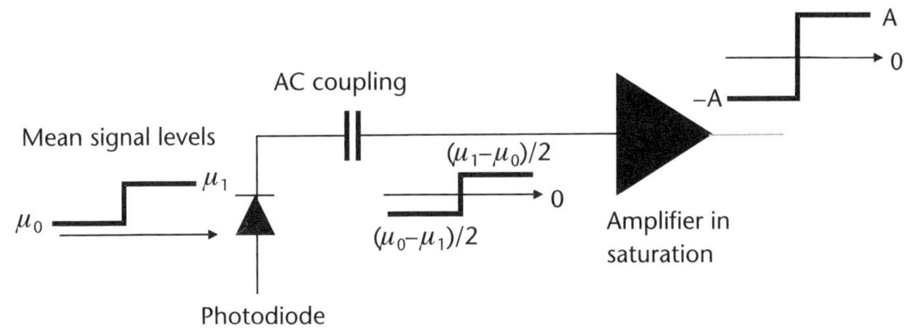

Figure 4.4 AC-coupled receiver.

signal level to zero and a saturated amplifier provides two levels of output signal, A and –A, for transmitted mark and spaces, respectively.

For this receiver, the threshold is set half-way between mark and space levels: $I_{th} = \frac{\mu_1 - \mu_0}{2}$ and the BER can be calculated defining two different Q factors, conditioned to the transmission of 1 and 0 bits:

$$Q|1 = \frac{\mu_1 - I_{th}}{\sigma_1} \quad Q|0 = \frac{I_{th} - \mu_0}{\sigma_0} \tag{4.13}$$

For equiprobable transmitted bits the BER is:

$$BER = \frac{1}{2} erfc\left(\frac{Q|1}{\sqrt{2}}\right) + \frac{1}{2} erfc\left(\frac{Q|0}{\sqrt{2}}\right) \tag{4.14}$$

Equation (4.10) do not apply to a multilevel signal like PAM4, but it is still possible to use the Q factor to estimate the symbol error rate (SER), assuming that most of the errors are due to a swap of two adjacent amplitude levels, and errors due to nonadjacent symbols can be neglected; and the decision threshold between two adjacent amplitude levels is set to the optimum value. For example, numerating the four levels of a PAM4 signal from 1 to 4, it is possible to define three Q factor values:

$$Q_{21} = \frac{\mu_2 - \mu_1}{\sigma_2 + \sigma_1} \quad Q_{32} = \frac{\mu_3 - \mu_2}{\sigma_3 + \sigma_2} \quad Q_{43} = \frac{\mu_4 - \mu_3}{\sigma_4 + \sigma_3} \tag{4.15}$$

Let n_{12} be the number of errors due to transmitted 2 symbols wrongly received as 1 symbols (with similar definitions holding for the other signal levels); n_E is the total number of errored symbols; and N the total number of transmitted symbols. Then, the following series of equalities can be written:

$$SER = \frac{n_E}{N} \sim \frac{n_{21} + n_{12} + n_{32} + n_{23} + n_{43} + n_{34}}{N}$$
$$= 2\frac{n_{21} + n_{32} + n_{43}}{N} = \frac{1}{2}\left(\frac{n_{21}}{\frac{N}{4}} + \frac{n_{32}}{\frac{N}{4}} + \frac{n_{43}}{\frac{N}{4}}\right)\backslash \quad (4.16)$$
$$= \frac{1}{4}\left(erfc\left(\frac{Q_{21}}{\sqrt{2}}\right) + erfc\left(\frac{Q_{32}}{\sqrt{2}}\right) + erfc\left(\frac{Q_{43}}{\sqrt{2}}\right)\right)$$

In (4.16) it is assumed that all symbols are equiprobable. The BER can be derived from the SER if the bit coding rule is known; for example, with a Gray code, adjacent symbols are coded in bit sequences that differ just in one bit.

4.3.3 Optical Modulation Amplitude

The numerator of the Q factor in (4.8) depends on the difference of the two signal amplitude levels that, in a direct detection optical communication system, are proportional to the received optical power levels, P_1 and P_0. The OMA is defined as:

$$OMA = P_1 - P_0 \quad (4.17)$$

The OMA can be related to the extinction ratio, r, defined as:

$$r = \frac{P_1}{P_0} \quad (4.18)$$

and to the average power level, P_{av}:

$$P_{av} = \frac{P_0 + P_1}{2} \quad (4.19)$$

$$OMA = 2 \cdot P_{av} \frac{r-1}{r+1} \quad (4.20)$$

The OMA is specified by the transceiver manufacturer that, such a way, has the design choice to guarantee a given performance either by acting on the average power or by acting on the extinction ratio. As for the Q factor, the OMA can be extended to multilevel modulation formats by defining it for each pair of signal levels.

4.3.4 Error Vector Magnitude

The EVM is a parameter used for modulation formats, such as QPSK and QAM, represented by a constellation of symbols in the complex plane, or I/Q plane. It is defined as the root mean square (rms) value of the Euclidian distance (i.e., the vector error) between the measured positions and the nominal positions of the symbols in the I/Q plane (Figure 4.5).

For a constellation with N symbols, the EVM obeys to the following equations:

$$EVM_n = \sqrt{I_{err,n}^2 + Q_{err,n}^2}$$
$$EVM_{rms} = \frac{\sqrt{\frac{1}{N}\sum_{n=1}^{N} EVM_n^2}}{A} \tag{4.21}$$

In (4.21), A is a normalization factor. Typically, it is the highest module among the symbols of the constellation or the average power of the symbols. For a noiseless signal, nominal and measured position collapse and $EVM_{rms}=0$.

4.3.5 Optical Signal-to-Noise Ratio

The OSNR is defined for an amplified optical link as:

$$OSNR = \frac{P}{N} \tag{4.22}$$

Where P is the signal optical power and N is the optical power resulting from the integration of the amplification noise power spectral density (psd), N_0, over an arbitrary reference bandwidth, B_0, called resolution bandwidth. The resolution bandwidth is chosen sufficiently small, so that the noise psd is approximately constant over it:

$$N \sim N_0 \cdot B_0 \tag{4.23}$$

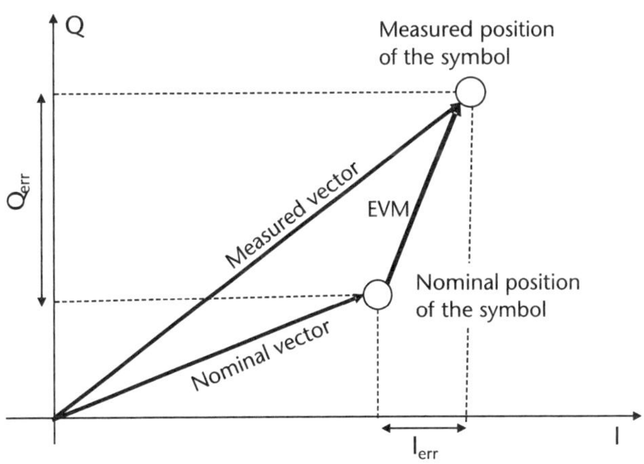

Figure 4.5 Visual representation of the EVM.

A typical choice for B_0 is 12.5 GHz, which approximately corresponds to 0.1 nm in C-band.

The amplification noise is mostly due to spontaneous photons emission and is called therefore amplified spontaneous emission (ASE) noise. The ASE light is unpolarized and its psd can be equally split between two orthogonal linear polarization axes, e_X and e_Y, and $N_{0\text{-}X}=N_{0\text{-}Y}=N_0/2$.

As the Q factor, the OSNR is often expressed in decibels:

$$OSNR_{dB} = 10 \cdot log_{10}(OSNR) \tag{4.24}$$

In an optical network, two amplified optical links, or optical network elements (ONEs), can be concatenated by means of an optical cross connect (OXC) or a reconfigurable optical add-drop multiplexer (ROADM). The OSNR of the concatenated link can be calculated from the individual OSNRs of the two links, $OSNR_1$ and $OSNR_2$, as shown in Figure 4.6.

In Figure 4.6, the network elements ONE1 and ONE2 generate the noise contributions N_1 and N_2. ONE2 amplifies, or attenuates, both signal and noise coming from ONE1 by a factor g_2. It is easy to verify from Figure 4.6 that the OSNR resulting from the cascade of the two network elements is:

$$OSNR = \left(OSNR_1^{-1} + OSNR_2^{-1}\right)^{-1} \tag{4.25}$$

Equation 4.25 can be applied iteratively to calculate the OSNR in a cascade of several network elements.

4.3.6 Using Different Penalty Definitions

Performance impairments due to noise or propagation distortions are modeled by allocating penalties on received power, *Q* factor or OSNR. There is no general rule to use one or other of these penalty definitions, but it depends on the nature of the problem. For example, a power penalty can be easily applied to a point to point unamplified link, as the path penalty in (4.5), where a new value of receiver sensitivity is calculated adding the penalty to the intrinsic receiver sensitivity:

$$Rx\,Sensitivity_{new} = Rx\,sensitivity + Power\,Penalty\ (dB) \tag{4.26}$$

A penalty on the *Q* factor is allocated either subtracting a term to the numerator of (4.8), or adding it to the denominator. The former method is equivalent to

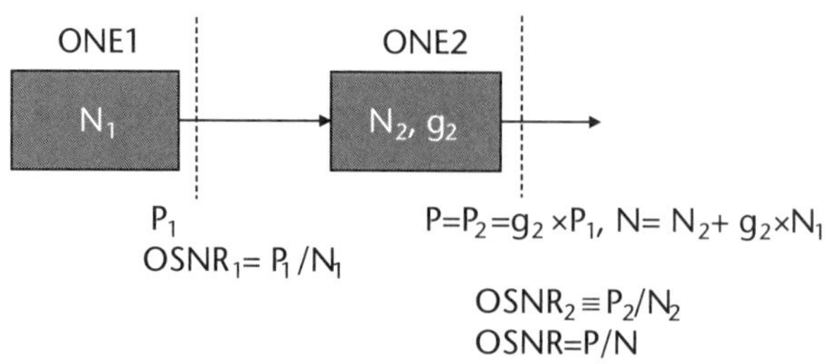

Figure 4.6 OSNR in a cascade of two optical network elements.

consider a smaller distance between the signal levels, or a smaller extinction ratio, and is also called eye closure penalty. With the second method, the penalty is modeled as a noise term, having an equivalent variance $\sigma_{p1,0}^2$, where the 1 or 0 subscripts refers to the value of the transmitted bit. The new standard deviation terms at the denominator of (4.8) are calculated as:

$$\sigma_{1,0\,new} = \sqrt{\sigma_{1,0}^2 + \sigma_{p\,1,0}^2} \tag{4.27}$$

In many practical situations, both σ_0 and σ_{p0} are negligible compared to σ_1. In these cases, the new value of Q factor can be calculated as:

$$Q_{new}^{-2} \sim \frac{\sigma_1^2 + \sigma_{p1}^2}{\left(\mu_1 - \mu_0\right)^2} \tag{4.28}$$

Defining:

$$Q_p = \frac{\mu_1 - \mu_0}{\sigma_{p1}} \tag{4.29}$$

Equation (4.28) can be rewritten as:

$$Q_{new}^{-2} = Q^{-2} + Q_p^{-2} \tag{4.30}$$

and the Q factor penalty can be defined as:

$$P_Q = \frac{Q^2}{Q_{new}^2} = 1 + Q^2 \cdot Q_p^{-2} \tag{4.31}$$

Note that, for a given value of σ_P, higher penalties are obtained for higher values of Q (i.e., lower BER).

The OSNR penalty is defined as for the Q factor: allocating a penalty to the numerator of (4.22) is equivalent to consider a power penalty; adding instead an equivalent noise term, N_p, to its denominator, the following penalty value is obtained:

$$P_{OSNR} = \frac{OSNR}{OSNR_{new}} = 1 + OSNR \cdot OSNR_p^{-1} \tag{4.32}$$

The meaning of the different terms in (4.32) can be deduced from the previous discussion.

4.4 Optical Receiver Model

In a direct detection optical receiver (Figure 4.7), a photodiode (PD) converts an input optical signal in an electrical current, called photocurrent.

The photocurrent, I, is proportional to the instantaneous incident optical power, $|E(t)|^2$ (see Figure 4.7):

$$I(t) = \mathfrak{R} \cdot |E(t)|^2 \tag{4.33}$$

The photodiode is insensitive to the phase of the incident optical signal, $\vartheta(t)$. The constant $\mathfrak{R}$, measured in amperes/watt, is called responsivity, and measures the efficiency of the conversion process. It is given by:

$$\mathfrak{R} = \eta \frac{q \cdot \lambda}{h \cdot c} \tag{4.34}$$

The quantum efficiency η is the percentage of incident photons which are converted into electrons, for example 75%, q is the electron charge ($1.60217662 \times 10^{-19}$ coulomb), λ is the optical carrier wavelength, h is Planck's constant ($6.62607004 \times 10^{-34}$ joule/s) and c is the speed light in vacuum (299792458 m/s). Two types of PD exist, photo-intrinsic diodes (PIN) and avalanche photodiodes (APD). They differ in the input power range, defined as the range a target value of BER (e.g., 10^{-12}), is guaranteed within. The lower bound of the input power range is the receiver sensitivity, while the upper bound is called overload. The APD power range is downshifted compared to a PIN, thanks to the gain of the avalanche multiplication effect, or multiplication factor, M. M depends on the reverse voltage applied to the APD and, through it, on the input optical power, because the photocurrent generates a voltage drop across the photodiode. The dependency of M by input optical power and reverse voltage is specified in the component datasheet; typical values are included between 5 and 10. Both PIN or an APD introduce two noise terms: Dark current and shot noise. The dark current is the current in absence of input optical power. The shot noise is due to the quantum nature of the photodetection process. Both dark noise and shot noise can be modeled as a Poisson random processes. In an APD, the avalanche multiplication process generates excess shot noise, modeled by an excess noise factor, F, which depends on M and on a constant, κ ($0<\kappa<1$, typically ~0.1):

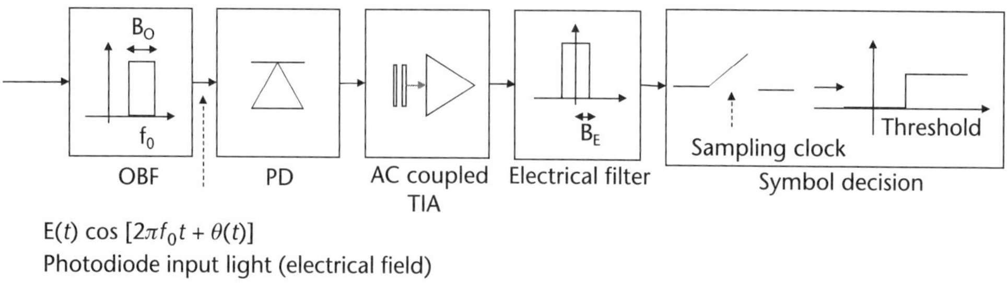

Figure 4.7 Optical receiver model.

$$F = \kappa \cdot M + (1-\kappa) \cdot \left(2 - \frac{1}{M}\right) \tag{4.35}$$

In the following, the light at the PD input will be modeled as an OOK modulated signal with bit rate Rb, extinction ratio *r*, and duty cycle *dc*, corrupted by additive relative intensity noise (RIN) and ASE noise. The duty cycle for an OOK signal is defined as the ratio of the time over which the signal equals the upper level, and the total transmission time. The RIN is caused by photon spontaneous emission in the laser and leads to amplitude fluctuations proportional to the laser output power. It is measured in dB/√Hz and has typical values from –120 to –140 dB/√Hz. Signal and noise are filtered by an optical bandpass filter with equivalent noise bandwidth B_o, centered at the optical carrier frequency. B_o is slightly higher than twice the modulation bandwidth, so to not introduce significant ISI. The PD is AC coupled to a transimpedance amplifier (TIA). The AC coupling eliminates the continuous current, which absorbs energy and carry no information. DC coupling is, however, required in the burst mode receivers used in Passive Optical Networks (PONs), which have to quickly recover input signals with sudden amplitude transitions. The TIA increases the receiver cut-off frequency, mitigating the effects of the PD intrinsic capacitance, but it also introduces thermal noise, modeled by the electrical noise figure, NF_e, and the transimpedance value, R. The combined frequency response of PD, TIA, and the other electrical components of the receiver is modeled by the electrical filter in Figure 4.7, having an equivalent noise bandwidth B_E, slightly wider than the Nyquist bandwidth 0.5·Rb. Typically, it is $B_E = 0.7 \cdot \text{Rb}$. Ideally, the filter shape should be matched to the signal pulse shape, but a filter widely adopted in practice is a fourth-order Bessel Thomson filter.

After filtering, the signal is sampled, and compared to the decision threshold; samples with an amplitude higher than the threshold are estimated as marks, and samples below the threshold are estimated as spaces. The BER resulting from the decision process is estimated using the Q factor, according to (4.10) or (4.14), for the optimal decision threshold or threshold set to zero, respectively. To calculate the Q factor, average and standard deviation of the received signal will be calculated. The discussion will be carried out for an OOK signal, but the same methodology applies to multilevel signals, calculating mean and variance for all signal levels. To consider both the cases of non-return-to-zero (NRZ) and return-to-zero (RZ) pulse shape for an OOK modulated signal, the average signal values for marks and spaces, P_1 and P_0, are calculated as function of average optical power, P_{av}, extinction ratio, *r*, and duty cycle, *dc*:

$$P_1 = \frac{r \cdot P_{av}}{1 + (r-1)\dfrac{dc}{2}} \qquad P_0 = \frac{P_{av}}{1 + (r-1)\dfrac{dc}{2}} \tag{4.36}$$

The average values at the numerator of the Q factor are:

$$\begin{aligned} \mu_1 &= \Re \cdot M(P_1 + P_{ASE}) + I_d \\ \mu_0 &= \Re \cdot M(P_0 + P_{ASE}) + I_d \end{aligned} \tag{4.37}$$

In (4.37), I_d is the dark current, specified in the PD datasheet. The dark current of an APD is the sum of two contributions: Surface leakage current I_{DS}, and bulk leakage current I_{DB} [8]:

$$I_d = I_{DS} + M \cdot I_{DB} \tag{4.38}$$

In optically amplified systems, the ASE noise contributes to the average values with the term P_{ASE}:

$$P_{ASE} = N_0 \cdot B_o \tag{4.39}$$

where N_0 is the ASE psd and B_o is the bandwidth of the optical filter. N_0 is calculated as:

$$N_0 = \frac{P_{av}}{OSNR \cdot B} \tag{4.40}$$

where B is the OSNR resolution bandwidth.

The Q factor denominator is the sum of several noise terms, calculated next:

The thermal noise variance is independent of the signal level and is given by (4.41)

$$\sigma^2_{therm} = \frac{4 \cdot k \cdot T \cdot NF_e \cdot B_E}{R} \tag{4.41}$$

In (4.41), k is Boltzmann's constant ($1.38 \cdot 10^{-23}$ J/K), T is the absolute temperature (e.g., 300°K), B_E is the electrical bandwidth (e.g., $0.7 \cdot R_b$), NF_e is the noise figure of the electrical frontend (e.g., 3 dB), and R is the transimpedance value (e.g., 50 Ω).

The shot noise variance is proportional to average optical power and dark current [8]:

$$\begin{aligned} \sigma^2_{shot\,1,0} &= \begin{pmatrix} 2 \cdot q \cdot \Re \cdot M^2 \cdot \left(P_{1,0} + P_{ASE}\right) \cdot \\ F + 2 \cdot q \cdot I_{DS} + 2 \cdot q \cdot M^2 \cdot I_{DB} \cdot F \end{pmatrix} \cdot B_E \quad APD \\ \sigma^2_{shot\,1,0} &= \left(2 \cdot q \cdot \Re \cdot \left(P_{1,0} + P_{ASE}\right) + 2 \cdot q \cdot I_d\right) \cdot B_E \quad PIN \end{aligned} \tag{4.42}$$

The RIN variance can be calculated as:

$$\sigma^2_{RIN\,1,0} = \Re^2 \cdot M^2 \cdot P^2_{1,0} \cdot rin \cdot B_E \tag{4.43}$$

where RIN is the laser relative intensity noise converted from dB/√Hz into linear units.

The ASE noise generates two variance terms, corresponding to signal-to-ASE beating (4.45) and ASE-ASE beating (4.44) in the PD:

$$\sigma_{sp-sp}^2 = \Re^2 \cdot M^2 \frac{N_0^2}{2}\left(B_o \cdot B_E - \frac{B_E^2}{2}\right) \tag{4.44}$$

In (4.44), it is assumed that N_0 is equally split between polarization states orthogonal and parallel to the signal polarization axis. The signal-ASE beating term is:

$$\sigma_{s-sp\,1,0}^2 = 2 \cdot \Re^2 \cdot M^2 \cdot P_{1,0} \cdot N_0 \cdot B_E \tag{4.45}$$

Finally, the standard deviation terms in the Q factor can be calculated as:

$$\sigma_{1,0} = \sqrt{\sigma_{therm}^2 + \sigma_{shot\,1,0}^2 + \sigma_{RIN\,1,0}^2 + \sigma_{sp-sp}^2 + \sigma_{s-sp\,1,0}^2} \tag{4.46}$$

Figure 4.8 compares all the terms calculated above for a PIN receiver with a transimpedance of 50Ω and an optical bandwidth of 0.4 nm. The input signal is a 10 Gbit/s NRZ signal with *dc*=1, *r*=10 dB and *RIN*=–130 dB/√Hz. Average signal power and OSNR are –15 dBm and 20 dB, respectively. Figure 4.8 shows that the signal-to-ASE beating is dominant in optically amplified systems.

4.5 Fiber Propagation Penalties

A signal propagating into fiber experiences not only attenuation but also linear and nonlinear, or signal power dependent, waveform distortion. Chromatic dispersion (CD) and polarization mode dispersion (PMD) are two linear propagation effects, caused by the dependency of the signal group velocity on frequency or polarization, respectively. Several nonlinear effects are caused by the dependence of the fiber effective refractive index on the signal power, a phenomenon known as Kerr effect. Self-phase modulation (SPM), cross-phase modulation (XPM), and four-wave mixing (FWM) belong to this category. Other nonlinear effects are due to stimulated

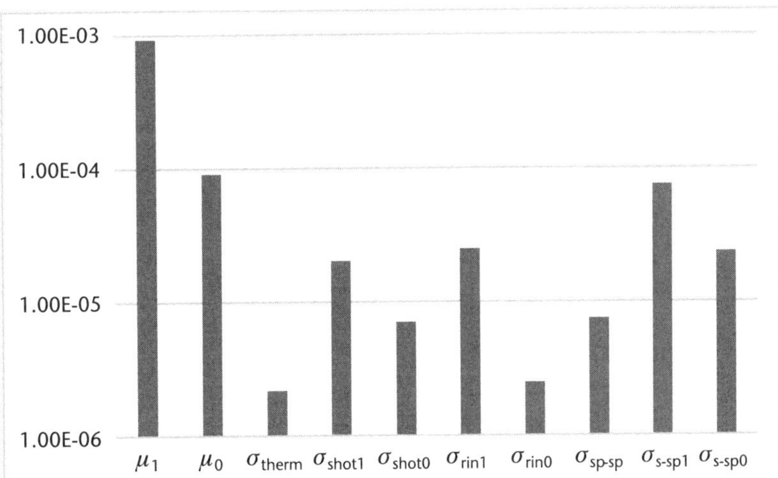

Figure 4.8 Example of average and standard deviation values for a 10-Gbit/s system.

Brillouin scattering (SBS) and stimulated Raman scattering (SRS) in fiber. In addition, bidirectional systems are affected by linear signal backscattering, or Rayleigh backscattering (RBS). The intent of this section is to provide optical network designers with simple recipes to estimate penalties and allocate margins for all these effects. A comprehensive analysis of the underlying physics is outside of the scope of this book.

4.5.1 Chromatic Dispersion

CD is the phenomenon whereby different spectral components of the same signal travel at different speed. This leads to pulse broadening during the propagation in fiber, having the twofold effect of decreasing the pulse peak power, since the pulse energy remains constant in lossless linear propagation regime, and causing ISI, since the pulse tails overrun the adjacent pulses. Simplified analytical models, considering both the effects, are proposed in [1] and [9]. A more accurate model, encompassing all factors that contribute to the CD penalty, such as actual pulse shape, bit pattern effects, presence of chirp, laser linewidth, and shape of the optical and electrical filters at the receiver, is described in [10]. For moderately high signal powers, an accurate model should also consider the SPM, which chirps the signal at the beginning of each fiber span [11]. To avoid all that complication, system designers often adopt an ultrasimplified approach, where: (a) the maximum acceptable value of accumulated CD, corresponding to the maximum acceptable penalty, is fixed by design (for example, it could be the accumulated dispersion leading to 2 dB of sensitivity penalty, which is about 800 ps/nm for an unchirped NRZ signal); (b) the curve giving the penalty versus the accumulated CD is upper bounded by a constant or a piecewise line, as in the example of Figure 4.9. For Q factor and OSNR penalties, reference values of OSNR or Q factor must be specified as well, according to (4.31) and (4.32). A reference BER is often provided instead of the Q factor, using (4.10). Usual values of reference BER are 10^{-12} for systems without FEC and 10^{-5} for systems using a Reed Solomon RS (255, 239) code, as in the OTN standard [4].

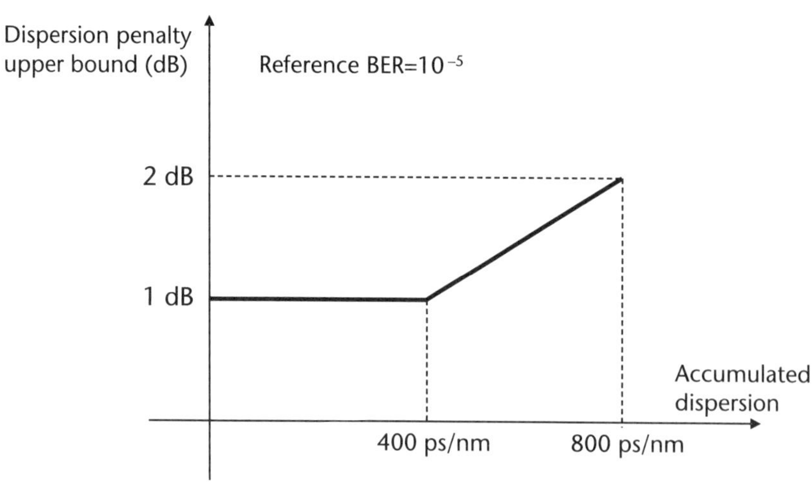

Figure 4.9 Example of dispersion penalty upper bound

As rule of thumb, if the bit rate is much higher than the laser linewidth, the accumulated dispersion at 1 dB penalty scales with the square of bit rate R_b and channel wavelength, λ [1]:

$$Accumulated\_Dispersion \cdot R_b^2 \cdot \lambda^2 \sim 282.548 \tag{4.47}$$

A link composed by N fiber spans, separated by optical amplifiers or OADMs, usually includes dispersion compensation modules (DCMs), which must be considered in the total accumulated CD [see (4.48)]:

$$Accumulated\,dispersion = \sum_{n=1}^{N} D_n \cdot L_n + \sum_{n=0}^{N} DCM_n \tag{4.48}$$

where L_n is the span length in kilometers, D_n is the span CD coefficient, measured in ps/(nm·km), and DCM_n, is the negative dispersion amount, in ps/nm, of the DCM placed after the nth fiber span. DCM_0 corresponds to a DCM placed before the first span, sometimes indicated as precompensation module.

The chromatic dispersion coefficient D depends on the wavelength and, in C-band, can be approximated as:

$$D(\lambda) = D(\bar{\lambda}) + S(\bar{\lambda}) \cdot (\lambda - \bar{\lambda}) \tag{4.49}$$

where $\bar{\lambda}$ is a reference wavelength and S is the dispersion slope in ps/(nm^2·km).

Dispersion coefficient and dispersion slope coefficient, referred to a wavelength of 1,546 nm, are reported in Table 4.2 for different types of fiber.

In Table 4.2, large effective area fiber (LEAF), enhanced LEAF (E-LEAF), True-Wave (TW) RS, plus and classical (clas), and Teralight are commercial names of fibers belonging to the family of the non -zero-dispersion shifted fibers (NZ-DSF). DSF and SMF stay for dispersion shifted fiber and single mode fiber.

Table 4.3 provides examples of the same parameters for DCMs designed to compensate the CD of a SMF span.

4.5.2 Polarization Mode Dispersion

Ideally, the transversal section of an optical fiber has a perfect circular symmetry, and its fundamental propagation mode HE_{11} can be considered as the superimposition of two orthogonal, linearly polarized modes having equal propagation

Table 4.2 Dispersion Coefficient and Slope for Different Types of Fiber

Parameter	*SMF*	*E-LEAF*	*LEAF*	*TW-RS*	*TW-plus*	*TW-clas*	*Teralight*	*Vascade*	*DSF*
Dispersion ps/nm/km	16.46	3.9	3.36	4.53	3.33	2.45	7.768	18.46	–0.325
Disp. Slop ps/nm^2/km e	0.06	0.084	0.11	0.043	0.068	0.068	0.058	0.06	0.074

Table 4.3 Characteristics of DCMs for SMF

Compensated SMF Fiber Length km	*Disp. ps/nm*	*Slope ps/nm²*	*Insert Loss dB*
5	–81	–0.29	1.9
10	–163	–0.59	2.1
20	–325	–1.17	2.8
30	–487	–1.76	3.4
40	–649	–2.34	4.2
50	–811	–2.92	4.9
60	–973	–3.51	5.6
70	–1136	–4.10	6.2
80	–1298	–4.68	6.9
90	–1460	–5.26	7.6
100	–1622	–5.85	8.3
110	–1785	–6.44	9
120	–1947	–7.02	9.7
130	–2110	–7.61	10.4
140	–2273	–8.20	11.1
150	–2435	–8.78	11.9
160	–2597	–9.36	12.4

constants [3]. In real fibers, the propagation constant is slightly different for the two modes, due to birefringence introduced randomly along the fiber length, induced by imperfections or stress. This is known as the PMD effect [12]. First and second order PMD are defined expanding the propagation constants of the two modes in Taylor series versus the frequency and calculating the difference between the two first order terms and the two second order terms, respectively [13]. It is possible to demonstrate that a pair of orthogonal linear polarization states (principal states) exists at the fiber input such that the output states remain orthogonal and their relative delay shows no wavelength dependence, to first order. The relative propagation delay of two signals aligned to the two axes is called differential group delay (DGD). Due to fluctuations of temperature, humidity, pressure, or other stress factors, the fiber birefringence exhibits a statistical behavior, which results into a Maxwell pdf of the DGD:

$$p(\Delta\tau) = \frac{32\Delta\tau^2}{\pi^2 \Delta\tau^3} e^{-\frac{4\Delta\tau^2}{\pi\Delta\tau^2}} \tag{4.50}$$

where $\Delta\tau$ is the DGD and $\langle\Delta\tau\rangle$ is its mean value, calculated as function of the fiber PMD coefficient, in ps/√km, and the fiber length, L:

$$\langle\Delta\tau\rangle = PMD \cdot \sqrt{L} \tag{4.51}$$

The maximum tolerated DGD is expressed in multiple units of the average DGD, $\langle\Delta\tau\rangle$. The probability of the instantaneous DGD exceeding the maximum

DGD is called outage probability and can be inferred from its Maxwellian statistics. Some examples are given in Table 4.4.

In contrast with the CD, the PMD coefficient has no strong dependence on the type of fiber but depends on the quality of the installed cables. Values of 0.5 ps/√km, or higher, were common in old fibers while modern fiber shave a PMD coefficient lower than 0.1 ps/√km.

For an NRZ signal, the first-order PMD penalty can allocated as a power penalty [14]. Various penalty models exist: A simple one, reported in (4.52), where R_b is the bit rate, and $\Delta\tau$ is the DGD, is compared with measured data in Figure 4.10.

$$P_{PMD} = \frac{1}{\cos\left(\frac{\pi}{2}\Delta\tau \cdot R_b\right)} \tag{4.52}$$

4.5.3 Chromatic and Polarization Mode Dispersion Tolerance of Direct Detection Modulation Formats

Different modulation formats experience different CD and PMD penalties, and not always can a closed form approximation be found for the combined penalty. The penalty can be obtained anyway by numerical simulation, plotting its level curves in a plane having the accumulated chromatic dispersion on the x-axis, and the differential group delay on the y-axis. Examples of direct detection modulation suitable for a mobile transport network are listed in Figure 4.11.

The top of Figure 4.11 shows the constellation of the received symbols. NRZ-OOK, RZ-OOK, duobinary and combined amplitude phase shift (CAPS) [15] are all binary signals requiring a single photodiode at the receiver, as shown at the bottom of the figure; NRZ and RZ differential phase shift keying (DPSK) are binary signals that require an optical interferometer with a delay offset equal to one bit time, T, between the two interferometer arms, followed by a balanced photodiode; the receiver of NRZ and RZ differential quadrature phase shift keying (DQPSK) duplicate this structure, adding a $\pi/4$ phase offset between the interferometer arms.

A modulation format similar to CAPS, not shown in Figure 4.11, is optical duobinary [16]. Both CAPS and duobinary use narrowband optical filtering to improve spectral efficiency and tolerance to the CD, compensating the ISI introduced by the filter by means of line coding.

Table 4.4 Outage Probability versus Maximum DGD

Ratio of Maximum to Mean DGD	*Probability of Exceeding Maximum*
3.0	$4.2 \cdot 10^{-5}$
3.5	$7.7 \cdot 10^{-7}$
4.0	$7.4 \cdot 10^{-9}$

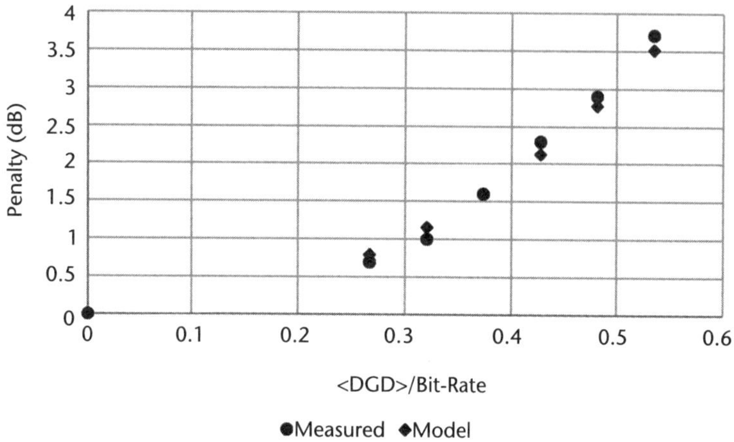

Figure 4.10 Measured PMD penalty versus model.

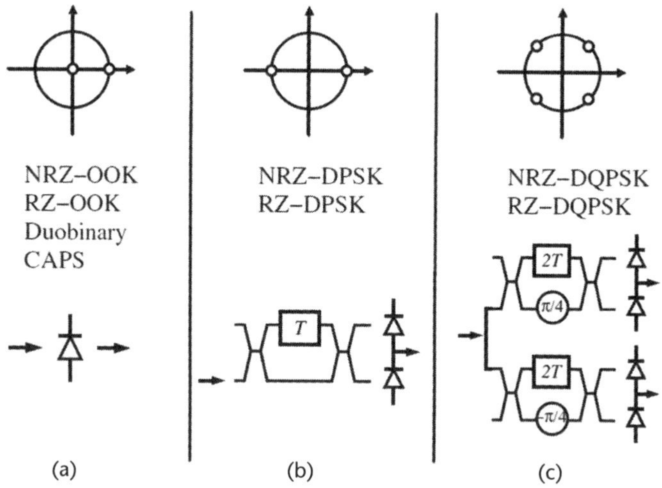

Figure 4.11 Examples of direct detection modulation formats.

Figure 4.12 reports level curves in the CD-PMD plane corresponding to a constant SNR, E_b/N_0, of 20 dB, for the aforementioned modulation formats. In Figure 4.12, the SNR is defined as the ratio between energy per bit E_b and ASE noise psd N_0. It is related to the OSNR by the equation:

$$\frac{E_b}{N_0} = 2\frac{B_0}{R_b} OSNR \tag{4.53}$$

where the factor 2 considers that N_0 is defined over two orthogonal polarization states and B_0 is the OSNR resolution bandwidth (see Section 4.3.5).

In Figure 4.12, both CD and PMD are expressed in units normalized to the bit rate. The PMD on the y-axis is given as ratio between mean DGD and bit time T_b=1/R_b, at an outage probability (OP) equal to 10^{-6}. The CD is expressed by means of the chromatic dispersion index γ, defined as:

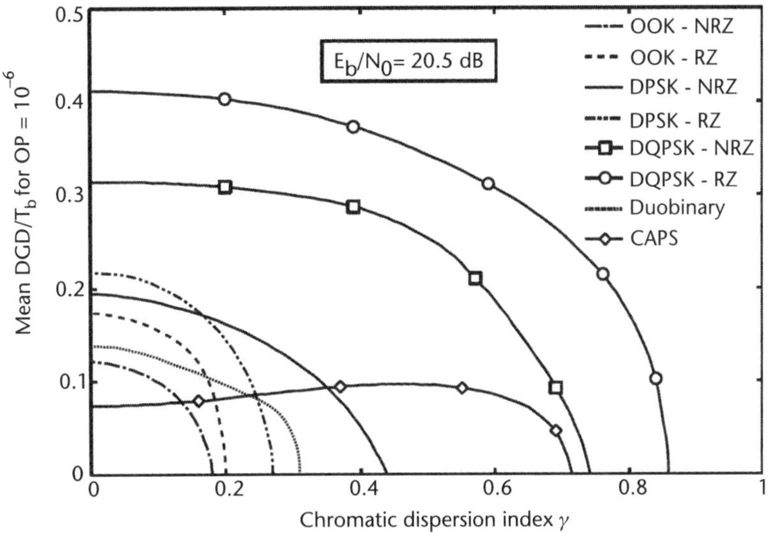

Figure 4.12 Chromatic and polarization mode dispersion tolerance of various direct detection modulation formats.

$$\gamma = \frac{\lambda^2 R_b^2 d}{\pi c} \tag{4.54}$$

where λ is the wavelength R_b, the bit rate d, the accumulated CD, and c, the speed of light. For example, γ=1 corresponds to 449 ps/nm at 10 Gbit/s and 28 ps/nm at 40 Gbit/s.

From Figure 4.12 it is evident that DQPSK exhibits by far the best performance, which is paid, however, by a considerable increase of implementation complexity compared to the other modulation formats. CAPS is very robust to the CD but it does not tolerate high values of PMD. DPSK is more tolerant than OOK to the OSNR, due to the higher distance between the constellation symbols.

4.5.4 Self-Phase Modulation

As its name suggests, SPM is a phase modulation induced on a signal propagating into fiber by power fluctuations of the signal itself. Since such a phase shift is proportional to the signal instantaneous power, the SPM is effective in the few kilometers of fiber after the transmitter or an optical amplifier, where it introduces a frequency chirp on the signal.

Due to the SPM, in an amplified optical link the received waveform depends on the position of the DCMs. The function that gives the accumulated chromatic dispersion versus the propagation distance is called dispersion map. An example of dispersion map with three fiber spans and three DCMs is shown in Figure 4.13. A fourth module, DCM_0, is present at the Tx to perform dispersion precompensation. Each DCM does not exactly reset the accumulated dispersion but leaves residual uncompensated dispersion at each span (undercompensation), including the Rx. In some dispersion maps, the accumulated dispersion is fully reset at the receiver (post-compensation).

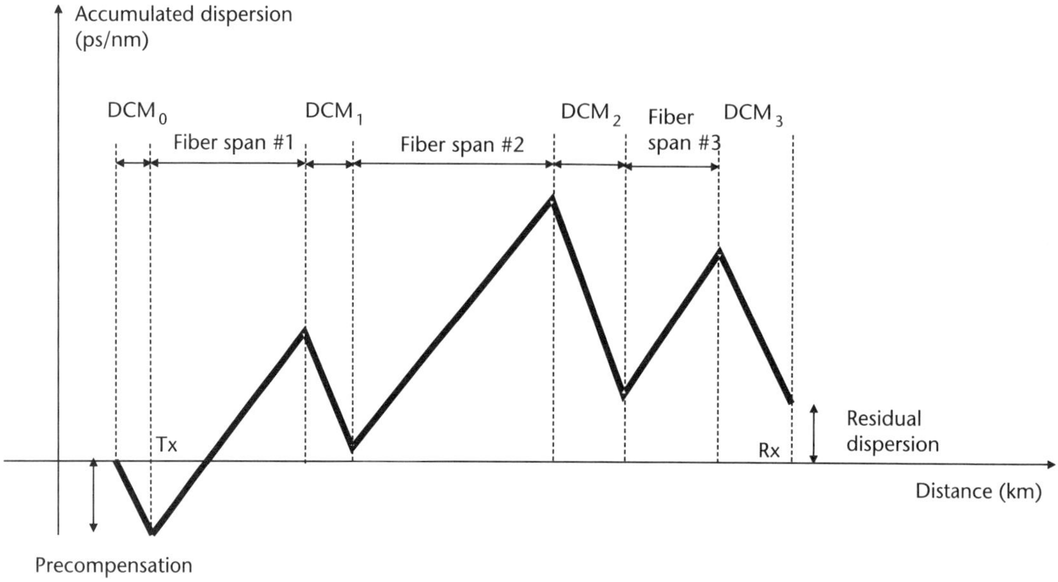

Figure 4.13 Example of dispersion map.

For an unchirped signal in linear propagation regime, the optimal residual dispersion at the receiver is zero and it does not matter how the dispersion map is configured. In a system affected by SPM, some precompensation is beneficial, as well as undercompensation at each fiber span. In fact, optimizing the dispersion map is one of the hardest design challenges in long-haul amplified systems [17]. Several analytical models have been proposed to quantify the SPM impact in a dispersive link, for example, [11] and [18], but numerical simulations remain the ultimate tool to estimate the received waveform and calculate the propagation penalty with high accuracy. In a RAN, the SPM is less important than in a long-haul system, due to the shorter distance, and number and position of DCMs are decided based on cost constraints rather than on performance optimization considerations. So, the SPM penalty can be included in the path penalty, adopting the ultrasimplified model of Figure 4.9. However, even in a RAN, the SPM limits the maximum channel power that can be launched in fiber and, because of the introduced chirp, the optimal residual dispersion at each span is different than zero. Figure 4.14 shows Q factor contour plots, in dB, versus residual dispersion per span, in ps/nm, and channel power, in dBm, for SMF transmission systems consisting of: A single span with an attenuation of 36 dB; 5 amplified spans, 25 dB each; 6 amplified spans, 27 dB each; 8 amplified spans, 22 dB each. The fiber parameters are: 0.23 dB/Km attenuation coefficient, 17 ps/(nm·Km) dispersion coefficient, 0.055 ps/(nm^2·Km) dispersion slope, 72 μm^2 effective area, 2.6 10^{-20} m^2/W nonlinear refractive index. Neither precompensation nor post-compensation are present and the DCMs after each span are all equal.

Optimal channel power and residual dispersion are reported in Table 4.5.

At a first glance, it seems odd that the optimal power is higher in the 6-span links than in the 5-span link. The reason is that in the 6-span link used as example the span attenuation is higher than in the 5-span link, so that the best performance is obtained with a higher channel power. This example shows that optimizing the

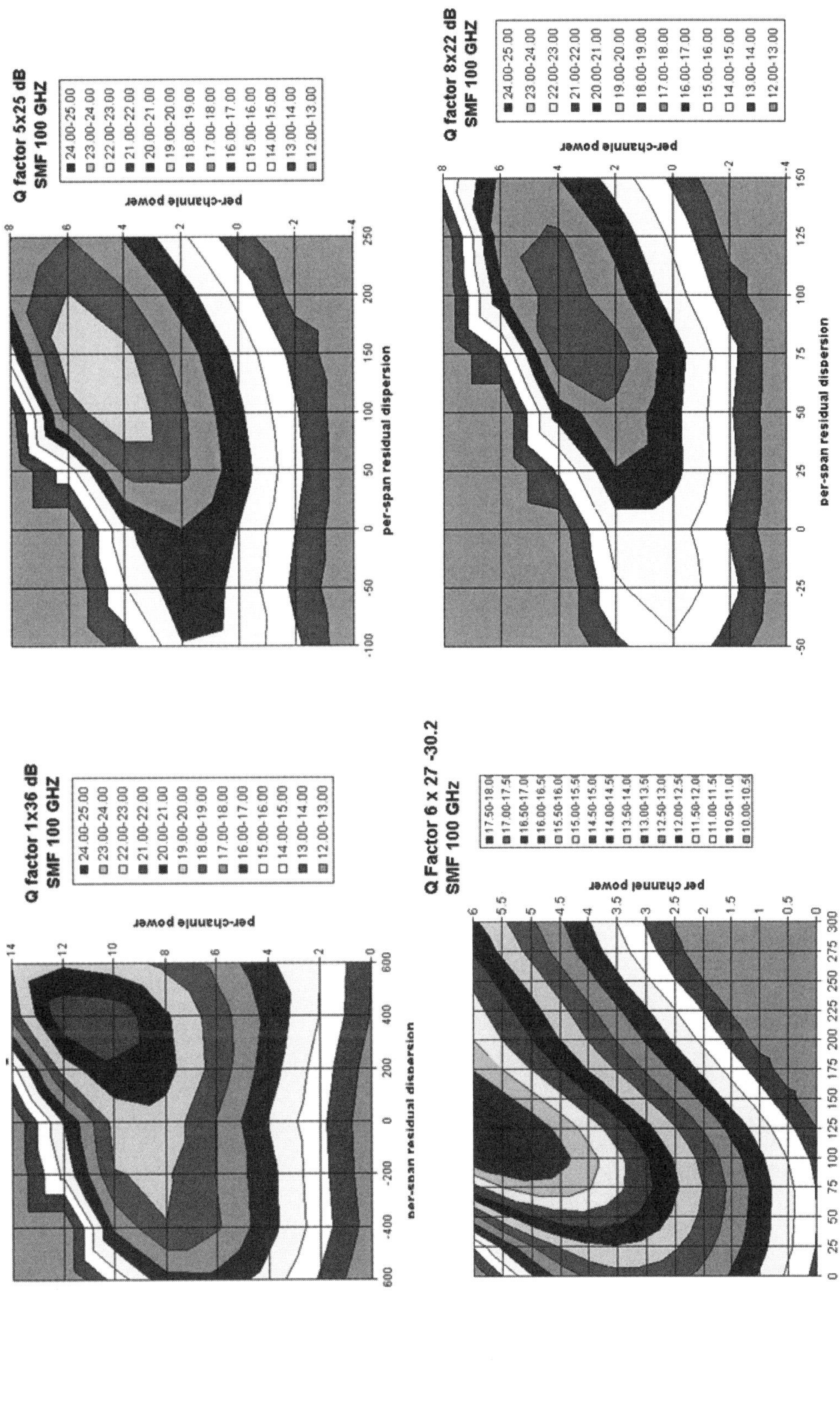

Figure 4.14 Q factor contour plots versus residual dispersion and channel power.

Table 4.5 Optimal Channel Power and Residual Dispersion in the Links of Figure 4.14

Number of Spans	*Max Channel Power dBm*	*Per Span Residual Dispersion ps/nm*
1	11	375
5	5	110
6	5.5	100
8	4	90

link design in presence of nonlinear effects is a complex task, requiring the tradeoff of several design variables.

4.5.5 Cross-Phase Modulation

Both SPM and XPM originate from the same power dependent term of the refractive index of the optical fiber. Due to XPM, intensity modulation in one optical carrier modulates the phase of another copropagating optical carrier. Since a direct detection receiver is insensitive to phase fluctuations, such phase crosstalk is not in itself cause of performance degradation. However, due to the chromatic dispersion, phase crosstalk is converted into intensity crosstalk, leading to a penalty. Several analytical models have been proposed to calculate the XPM penalty. The model presented in the following, based on [19], considers XPM as an additional noise term in the Q factor. In a DWDM system with M channels, the XPM variance, $\sigma^2_{XPM,\,j}$, for the jth channel, transmitted with an optical power P_j, is obtained summing up the contributions due to all other copropagating channels (the photodiode responsivity is omitted in the following equations for sake of simplicity):

$$\sigma^2_{XPM,j} = P_j^2 \sum_{\substack{k=1 \\ k \neq j}}^{M} \sigma^2_{XPM,jk} \tag{4.55}$$

The variances $\sigma^2_{xpm,\,jk}$, are obtained integrating the XPM power spectral density over the receiver bandwidth, B_E, assuming that the signal shape does not significantly change during the propagation over the optical link, consisting of N fiber spans:

$$\sigma^2_{XPM,jk} = \int_0^{B_E} \left|H_{jk}(f)\right|^2 \cdot S_k(f)\,df \tag{4.56}$$

where $S_k(f)$ is the power spectral density *of the power* of the kth channel and $H_{jk}(f)$ is the XPM transfer function [19]:

$$H_{jk}(\omega = 2\pi f) = \sum_{n=1}^{N} \left\{ 4 \cdot \gamma_n \cdot e^{i\omega \sum_{l=1}^{n-1} w_{jk}^{(l)} L_l} \cdot \frac{\sin\left[-\frac{\omega^2}{2} \sum_{h=n}^{N} \frac{\lambda^2}{2\pi c} \cdot u_h \right]}{\alpha_n - i\omega \cdot w_{jk}^{(n)}} \right\} \tag{4.57}$$

Where:

- N is the number of spans, including DCFs, considered as further spans;
- γ_n is the nonlinear coefficient of the span n ($1 \le n \le N$): $\gamma = 2\pi n_2 / A_{eff}$, where $n_2 = 2.6 \cdot 10^{-20}$ m^2/W is the nonlinear refractive index and A_{eff} is the fiber effective area;
- $w^{(n)}_{j,k}$ is the walk-off coefficient between channel j and channel k, at the span n: $w^{(n)}_{j,k} = D^{(n)} \cdot \lambda_{jk}$;
- $D^{(n)}$ is the chromatic dispersion coefficient of the span n;
- λ_{jk} is the wavelength separation between channel j and channel k;
- L_n is the length of the span n;
- λ is the reference wavelength at which the dispersion coefficient is calculated;
- α_n is the attenuation coefficient of the span n;
- u_h is the dispersion accumulated from the beginning of span h to the end of the link.

Equation (4.57) is derived solving the nonlinear Schrödinger equation that rules the propagation in fiber with an input signal, called pump, amplitude modulated by a tone at frequency ω, propagating together with a continuous wave signal, called probe. The model assumes that the interaction between SPM and XPM is negligible and pretends they act independently. This approximation is valid as long as the pump signal waveform is not appreciably changed by SPM. Furthermore, the intensity fluctuations of the probe, before it reaches the end of the system, are considered to be negligible. This model is valid when the XPM induces small perturbations from pump to probe. To simplify the model, making it a practical link design tool, the further approximation is made that the XPM contributions introduced at each span add up incoherently. For a single hop link (N=1), from (4.57) it is obtained:

$$\left|H_{jk}(\omega)\right|^2 = 16 \cdot \gamma^2 \cdot \frac{\sin^2\left[-\frac{\omega^2}{2} \frac{\lambda^2}{2\pi c} u \right]}{\alpha^2 + \omega^2 \cdot w_{j,k}^2} \tag{4.58}$$

An upper bound of the variance $\sigma^2_{XPM,\, jk}$, is calculated by applying Schwarz' inequality to (4.56).

$$\sigma^2_{XPM,jk} \le \sqrt{\int_0^{B_E} \left|H_{jk}(f)\right|^4 df} \cdot \sqrt{\int_0^{B_E} S_k^2(f) df} \tag{4.59}$$

For an NRZ signal, it is

$$S(f)=\frac{P^2}{R_b}\frac{\sin^2\left(\pi\frac{f}{R_b}\right)}{\left(\pi\frac{f}{R_b}\right)^2}$$

where P is the average signal power at fiber input. If B_E=R_b, it is $\int_0^{B_E} S_k^2(f)df \sim 0.332R_b$. Defining

$$I(x_0,x_1)=\int_0^{x_1}\frac{\sin^4\left(\frac{x^2}{x_0^2}\right)}{\left(1+x^2\right)^2}dx \tag{4.60}$$

with

$$x_0=\frac{w_{j,k}}{\alpha}\sqrt{\frac{4\pi c}{\lambda^2 u}} \quad x_1=\frac{2\pi\cdot R_b\cdot w_{j,k}}{\alpha} \tag{4.61}$$

It is

$$\sigma^2_{XPM,jk}\leq 9.2\frac{\gamma^2\cdot P_k^2}{\sqrt{2\pi\cdot w_{jk}\cdot\alpha^3\cdot R_b}}\sqrt{I(x_0,x_1)} \tag{4.62}$$

In a multispan link, it is sufficient to add up the individual variances calculated for each span, taking as accumulated dispersion value, u, the dispersion accumulated from the beginning of the current span to the end of the link. Thus, for the first span u is the residual dispersion at the end of the link, for the second span it is the dispersion at the end of the link minus the residual dispersion at the end of the first span, and so on.

4.5.6 Four-Wave Mixing

As SPM and XPM, FWM is an effect caused by the nonlinear refractive index of the fiber. Due to FWM, three optical carriers having frequencies f_p, f_q, and f_r, (with f_r different from both f_p and f_q), generate a fourth optical carrier at frequency $f_{FWM} = f_p + f_q - f_r$. If f_p and f_q are equal, the resulting FWM optical carrier is called degenerate FWM term. In an optical communication system, FWM transfers power from data channels to new frequencies, an effect which is known as power depletion. In DWDM systems, such new frequencies can coincide with other data channels, generating interference, which is a more serious issue than the depletion. In fact, the power depletion penalty is usually negligible (indeed the undepleted channel approximation is the basis of all FWM analytical models) while significant penalty is introduced by interference. For equally spaced channel frequencies, the number of FWM terms rapidly increases with the number, C, of DWDM channels, as 1/2 $(C^3 - C^2)$. The FWM power is calculated in [20] for a DWDM link consisting of M optically amplified spans. Each span is constituted by N segments of fiber, of equal length L_0. Within each segment, the zero-dispersion wavelength is constant. The model assumes continuous wave (CW) copolarized channels. Under these assumptions, the FWM obeys to the following set of equations:

$$\begin{aligned}
f_{\mathrm{FWM}} &= f_p + f_q - f_r \quad p,q \neq r \\
P_{\mathrm{FWM}}(p,q,r) &= (d \cdot \gamma \cdot L_e)^2 P_p \cdot P_p \cdot P_r \cdot e^{-\alpha L} \cdot \eta \\
\eta &= \alpha^2 \cdot \left| \sum_{m=1}^{M} e^{i\sum_{1}^{m-1}\Delta\phi_k} \cdot \sum_{n=1}^{N} e^{\sum_{j=1}^{n-1}(-\alpha + i\Delta\beta_{mj})L_0} \cdot \frac{1 - e^{(-\alpha + i\Delta\beta_{mn})L_0}}{\alpha - i\Delta\beta_{mn}} \right|^2 \\
\Delta\beta_{mn} &= -\frac{\pi\lambda^4}{c^2}(f_p - f_r)\cdot(f_q - f_r)\cdot S \cdot \left[(f_p - f_{0,m}) + (f_q - f_{0,m})\right] \\
\Delta\phi_m &= \sum_{n=1}^{N} \Delta\beta_{mn} \cdot L_0
\end{aligned} \tag{4.63}$$

where:

- f_p, f_q, f_r are the channels frequencies;
- P_p, P_q, P_r are the channels powers;
- α is the fiber attenuation coefficient;
- $L = N \cdot L_0$ is the span length;
- γ is the nonlinear fiber coefficient;
- L_e is the effective span length, $L_e = \dfrac{1 - e^{-\alpha L}}{\alpha}$;
- c is the light speed;
- $f_{0,mn}$ is the zero-dispersion frequency of the nth segment of the span m;
- $\lambda = 2 \cdot c/(f_p + f_q)$;
- S is the dispersion slope of the fiber, at the wavelength λ;
- d is the degeneracy factor, equal to 1 if $p=r$ and equal to 2 if $p \neq r$.

In a DWDM system, several FWM terms have the same frequency. The FWM power at that frequency should be calculated by adding up the electric fields with their relative phases, which are generally unknown, and, then, squaring the sum. A generally accepted approximation is to sum up the optical powers instead of the electric fields. According to this approximation, in a DWDM systems with C equally spaced channels, the total power interfering with the kth channel is:

$$P_{FWM-TOT}(f_k) = \sum_{\substack{p,q,r=1 \\ p+q-r=k \\ p\geq q \quad p\neq r \quad q\neq r}}^{C} P_{FWM}(p,q,r) \tag{4.64}$$

The parameter η in (4.63) is called FWM efficiency: It is calculated by adding the FWM terms generated at each amplified span, with their phases $\Delta\phi_k$. Since those phases are unknown, a practical approximation is summing up the powers instead of the electric fields (4.65):

$$\eta \sim \alpha^2 \cdot \sum_{m=1}^{M} \left| \sum_{n=1}^{N} e^{\sum_{j=1}^{n-1}(-\alpha+i\Delta\beta_{mj})L_0} \cdot \frac{1-e^{(-\alpha+i\Delta\beta_{mn})L_0}}{\alpha - i\Delta\beta_{mn}} \right|^2 \tag{4.65}$$

All above equations are complicated, making difficult understanding how the FWM scales with system design variables like channel power and fiber parameters. For this purpose, examples of simple system configurations are considered in the following. Putting $P_p = P_q = P_r = P$, it can be easily verified that $P_{\text{FWM}} \propto P^3$ and $P/P_{\text{FWM}} \propto P^{-2}$, meaning that even a small channel power increase generates significant FWM penalty. An interesting case to analyze is $N = 1$, $M = 1$, and $L >> Le$, so that:

$$\eta \sim \frac{\alpha^2}{\alpha^2 + \Delta\beta^2} \tag{4.66}$$

The highest FWM efficiency, $\eta \sim 1$, is obtained when the zero-dispersion wavelength is inside the DWDM channels spectrum, so that $\Delta\beta \sim 0$. This is the case of dispersion shifted fibers (DSF), which are no longer used due to this issue. When the propagation constant mismatch $\Delta\beta$ is higher than the attenuation coefficient α, as in SMF systems operating in the C band, it is:

$$\eta \sim \frac{\alpha^2}{\Delta\beta^2} \propto \frac{1}{D^2 \cdot \Delta\lambda^4} \tag{4.67}$$

where D is the CD coefficient and $\Delta\lambda$ is the channel wavelength spacing. Equation (4.67) shows that the CD kills the FWM, which is indeed negligible in commercial DWDM systems with 50- or 100-GHz frequency spaced channels. However, since the FWM scales with the fourth power of the channel spacing, it can become significant at lower spacings.

Some care must be taken in applying the mathematical model of (4.63). Two channels with orthogonal polarization do not generate any FWM term, hence the assumption of all copolarized channels leading to overestimate the FWM power in a real system, where there is no polarization control. Experiments with sixteen 100-GHz spaced NRZ DWDM channels, with 3 dBm of channel power transmitted in fiber over a single 66.5-km DSF span, having zero-dispersion wavelength positioned close to the central frequency of the DWDM spectrum, showed 3 dB of sensitivity penalty compared to an equal length SMF span and further 4.5-dB penalty with polarization aligned channels compared to the case of random polarization. Another factor to consider is the assumption of CW channels instead of modulated ones. At symbol rates of common use, the FWM response can be considered instantaneous compared to the symbol time: A rigorous approach should consider in (4.63) all possible combinations of channel power corresponding to the modulation symbol levels on the three channels generating the FWM. For example, the OOK modulation leads to eight different nondegenerate terms, listed in Table 4.6, and four degenerate terms, listed in Table 4.7.

Using Table 4.6, it is easy to prove that for nondegenerate FWM terms in a OOK modulated system, it is still possible to adopt a CW approach, calculating the average channel power, P_{av}:

$$P_{FWM} \propto 1/8\begin{pmatrix} P|0^3 + P|0^2 \cdot P|1 + P|0^2 \cdot P|1 + P|0 \cdot P|1^2 + \\ P|0^2 \cdot P|1 + P|0 \cdot |1^2\, P|0 \cdot P|1^2 + P|1^3 \end{pmatrix}$$
$$= 1/8\left(P|0^3 + 3 \cdot P|0^2 \cdot P|1 + 3 \cdot P|0 \cdot P|1^2 + P|1^3\right)$$
$$= 1/8\left(P|0 + P|1\right)^3 = P_{av}^3$$

A similar analysis can be performed for degenerate terms, for which $P_{FWM} \propto 1/4 \cdot (P|0^3 + P|0^2 \cdot P|1 + P|0 \cdot P|1^2 + P|1^3)$. This time, there is no immediate relationship between average channel power and FWM power but, for high extinction ratios, the FWM power is proportional to $1/4 \cdot P|1^3 \propto 2\, P_{av}{}^3$. Thus, it is still possible continuing to use the CW model in (4.63), using a constant degeneracy factor equal to two also for degenerate terms.

The FWM power can be considered as an additional noise term in the OSNR (4.22) and the penalty calculated as an OSNR penalty (Section 4.3.6). Alternatively, a Q-factor penalty can be calculated according to [21]. This model is valid for SMF or NZ-DSF, where the FWM efficiency is dominated by the CD:

$$BER = \frac{1}{2}\left[Q\left(Q_0 \frac{2 - 3kS}{\sqrt{4 - 3kS}}\right) + Q(Q_0)\right] \tag{4.68}$$

Table 4.6 FWM Nondegenerate Terms with OOK Channels

p	*q*	*r*	*Probability*	$P_{FWM} \propto$
0	0	0	1/8	$P|0^3$
0	0	1	1/8	$P|0^2 \cdot P|1$
0	1	0	1/8	$P|0^2 \cdot P|1$
0	1	1	1/8	$P|0 \cdot P|1^2$
1	0	0	1/8	$P|0^2 \cdot P|1$
1	0	1	1/8	$P|0 \cdot P|1^2$
1	1	0	1/8	$P|0 \cdot P|1^2$
1	1	1	1/8	$P|1^3$

Table 4.7 FWM Degenerate Terms with OOK Channels

$p=q$	r	*Probability*	$P_{FWM} \propto$
0	0	1/4	$P\vert 0^3$
0	1	1/4	$P\vert 0^2 \cdot P\vert 1$
1	0	1/4	$P\vert 0 P\vert 1^2$
1	1	1/4	$P\vert 1^3$

where

- $Q(x) = \frac{1}{\sqrt{2\pi}} \int_x^\infty e^{-\frac{\xi^2}{2}} d\xi$;
- Q_0 is the Q factor in absence of FWM;
- $k = 100 \frac{\gamma \cdot P}{|D| \cdot \Delta f^2}$, where γ is the fiber nonlinear coefficient, P is the average channel power, D is the chromatic dispersion coefficient, and Δf is the channels frequency spacing;
- S is a random variable modeling the FWM accumulation in a multispan system: In an M span system, its pdf is
- $$f_M\left(\frac{S}{\sigma_1}\right) = \frac{2\sqrt{2}e^{-\sqrt{2}|S|}}{(M-1)!2^M} \sum_{m=1}^{M} \frac{(M+m-2)!\left(\sqrt{2}|S|\right)^{M-m}}{2^m (m-1)!(M-m)!} \tag{4.69}$$

The values of S not exceeded with a 99% probability are reported in Table 4.8. They can be used in (4.68) to estimate the BER in presence of FWM.

4.6 Stimulated Raman Scattering

SRS is a nonlinear effect occurring in optical fiber when one photon, called pump photon, is converted into a lower energy photon, and the difference of energy is absorbed by quantized lattice vibration, or phonon. In DWDM systems, SRS leads to power transfer from lower wavelength channels, which act as pump, to higher wavelength channels, leading to channel depletion. The effect is visible as an approximately linear tilt of the DWDM spectrum at the fiber output [22]. In modulated systems, statistical variations of the channel power are superimposed to the power tilt. In a DWDM system with N OOK channels, the power depletion of the shortest wavelength channel can be modeled as a Gaussian distributed random variable, having average, η_{SRS}, and variance, σ^2_{SRS}, reported in the following [23]:

Table 4.8

M	1	5	10	15	20	25	30	35	40
S	3.3	6.2	8.4	10.2	11.7	13	14.2	15.3	16.3

$$\eta_{SRS} = K\frac{N\cdot(N-1)}{4}$$
$$\sigma^2_{SRS} = K^2\frac{N\cdot(N-1)\cdot(2N-1)}{24} \tag{4.70}$$

It is

$$K = \frac{g'_R\Delta f\cdot P_{ave}\cdot L_{eff}}{A_{eff}} \tag{4.71}$$

where P_{ave} is the average power of the lowest frequency channel, Δf is the channel frequency spacing, g'_R is the Raman gain slope (~ $7\cdot10^{-14}$ $W^{-1}m/1.5\ 10^{13}$ Hz), and A_{eff} is the fiber effective area.

In presence of CD, the model in [23] is modified to take into account of the walk-off between bits transmitted over different channels. Two bits transmitted over two $n\Delta f$ frequency spaced channels interact over a finite distance L_w/n, where the walk-off distance L_w is defined as

$$L_w = \frac{1}{R_b\cdot D\cdot\Delta f} \tag{4.72}$$

In (4.72), R_b is the bit-rate, D is CD coefficient, and $\Delta\lambda$ is the channel wavelength spacing. Making the approximation that SRS is only active over the fiber effective length, L_{eff}, one bit on one of the two channels will interact with approximately nL_{eff}/L_w bits on the other channel. Moreover, due to reduced interaction length in presence of walk-off, the power depletion will decrease of a factor L_w/L_{eff}, independent on n. Hence, power depletion mean value and variance can be calculated as in (4.73).

$$\eta_{SRS} = \sum_{n=0}^{N-1}\frac{1}{2}aK\frac{nL_{eff}}{L_w} = K\frac{N\cdot(N-1)}{4}$$
$$\sigma^2_{SRS} = \sum_{n=0}^{N-1}\left(\frac{1}{2}aK\right)^2\frac{nL_{eff}}{L_w} = K^2\frac{L_w}{L_{eff}}\frac{N\cdot(N-1)}{8} \tag{4.73}$$

The reason for the factor 1/2 in (4.73) is that zero bits do not provide any contribution. Comparing (4.73) with (4.70), it can be noted that in presence of CD, the average power depletion does not change, but the variance is significantly reduced. The analysis can be extended to multiple dispersion compensated spans, for which the local chromatic dispersion coefficient D must be replaced by an average dispersion coefficient, obtained dividing the residual dispersion, after each span, by the span length. The resulting equations for mean value and variance are quite complicated and are not reported here. In the ideal case where every bit interacts with different bits at each span (high walk-off), the variance grows linearly with the number of spans. The worst case is the case of perfect dispersion compensation, when the same bits interact at each span, and the variance grows quadratically with the number of spans.

4.6.1 Stimulated Brillouin Scattering

SBS is caused by high intensity light launched into fiber (pump signal) that causes lattice vibrations. Such lattice vibrations generate sound waves resulting in longitudinal variations of the refractive index, like in a fiber grating. As in a grating, the Bragg condition for constructive interference occurs in the direction opposite to that of the pump signal, generating backscattered light, known as Stokes' wave, frequency down-shifted compared to the signal light of an amount equal to the acoustic wave frequency. The backscattered power suddenly increases when the input power exceeds a threshold, conventionally defined as the input pump power at which back-scattered and output pump power are equal. For a CW light source, the SBS threshold is [24]:

$$P_{thr}^{CW} = 21\frac{K \cdot A_{eff}}{g_B \cdot L_{eff}} \cdot \frac{\Delta\nu_p + \Delta\nu_B}{\Delta\nu_B} \tag{4.74}$$

where:

- K is the number of independent polarization states of the input light, equal to 1 or 2;
- A_{eff} is the fiber-effective area;
- g_B is the Brillouin gain (typically, $g_B = 5 \times 10^{-11}$ m/W);
- L_{eff} is the fiber effective length;
- $\Delta\nu_p$ is the full width half maximum (FWHM) linewidth of the light source (in the following, a Lorentzian spectrum is assumed, with $\Delta\nu_p = 20$ MHz);
- $\Delta\nu_B$ is the FWHM bandwidth of the Stokes wave (typically $\Delta\nu_B = 100$ MHz).

For an OOK signal with a bit rate R_b, the threshold becomes:

$$P_{thr}^{OOK} = \frac{P_{thr}^{CW}}{1 - \frac{R_b}{2\Delta_B}\left(1 - e^{-\frac{\Delta_B}{R_b}}\right)} \tag{4.75}$$

At usual bit rates it is $\Delta\nu_B \ll R_b$, so that $P_{thr}^{OOK} \sim 2 \cdot P_{thr}^{CW}$. According to (4.75), the threshold is 10 dBm for an OOK signal transmitted into a SMF with 80 μm^2 effective area and 0.22 dB/km attenuation coefficient. A way to increase the threshold above this value is frequency modulating (dithering) the optical carrier with a tone at frequency f_{dith}, introducing a maximum frequency deviation ΔF, so that:

$$P_{thr}^{CW} \sim 21\frac{K \cdot A_{eff}}{g_B \cdot L_{eff}} \cdot \frac{\Delta\nu_p + \Delta\nu_B + 2(\Delta F + f_{dith})}{\Delta\nu_B} \tag{4.76}$$

The modulating frequency f_{dith} is much lower than the bit-rate, so that no additional CD penalty is introduced. Moreover, f_{dith}^{-1} must be lower than the propagation time through the fiber effective length, so that the spectral broadening effect can be fully exploited. Finally, in an amplified system, f_{dith}^{-1} must be lower than a few ms, which is the typical EDFA response times. Value of f_{dith} of around 20 kHz meet all these conditions. With dithering, the SBS penalty become negligible. It is also negligible in carrier suppressed modulation formats, due to the small portion of signal power falling over the narrow SBS bandwidth (about 100 MHz).

4.7 Rayleigh Backscattering

Even in linear propagation regime, light propagating into fiber experiences backscattering due to material imperfections. The effect is known as Rayleigh Back Scattering (RBS). The loss of power of the forward propagating signal due to Rayleigh backscattered light is taken into account in the fiber attenuation coefficient. It is proportional to λ^{-4}, making it the most significant loss contribution at low wavelengths (see Figure 4.3). In bidirectional systems, the backscattered light interferes with the backward propagating signal, causing crosstalk, which is especially detrimental in systems where the spectra of the downstream signal and the upstream signal overlap, a condition known as interferometric crosstalk. A mathematical derivation of the statistics of the backscattered light can be found in [25]. In a fiber with attenuation coefficient α and length L, the fraction, R, of backscattered power on the transmitted power is:

$$R = \alpha_S \cdot S \cdot \frac{1 - e^{-2\alpha L}}{2\alpha} \tag{4.77}$$

where α_S is the Rayleigh attenuation coefficient and the recapture coefficient S is the fraction of backscattered power in the direction of the forward propagating signal. R is plot versus distance in Figure 4.15, with $S = 0.001$, $\alpha_S = 3.2 \cdot 10^{-2}$, and $\alpha = 0.2$ dB/km.

R tends to saturate for long distances, as intuitively explained by the fact that light backscattered in the initial kilometers of fiber is less attenuated and provides a higher contribution to the total backscattered light.

In a direct detection receiver, if the laser at the transmitter has a Lorentzian shape with a FWHM bandwidth Δf, the backscattered power after the photodetection is:

$$\sigma_{RBS}^2 = P_0 \cdot R \cdot \frac{1}{\pi} \tan^{-1} \frac{B_E}{\Delta f}$$

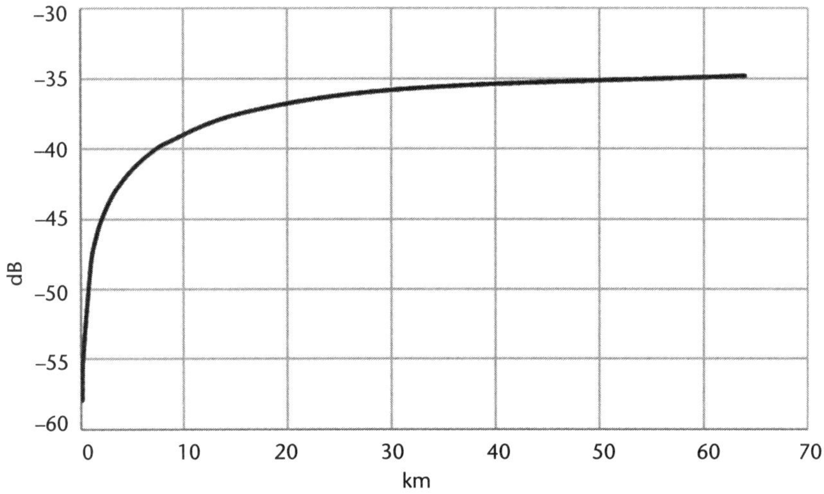

Figure 4.15 Rayleigh backscattered light versus distance.

where P_0 is the transmitted power and B_E is the receiver bandwidth. In systems where downstream and upstream signals are transmitted over different wavelengths, σ^2_{RBS} can be added as a further noise term in the Q factor. In the case of interferometric crosstalk, expressions for the penalty can be found in [1]. Here it is reported the penalty, in decibels, for a Gaussian interfering signal and a PIN receiver, at the average decision threshold (4.79) and at the optimal decision threshold (4.79).

$$P = -5 \cdot log_{10}\left[1 - 4 \cdot R \cdot Q^2 \frac{1 + r^{-1}}{\left(1 - r^{-1}\right)^2}\right] \tag{4.78}$$

$$P = -5 \cdot log_{10}\left[1 - 2 \cdot R \cdot Q^2 \left(\frac{r+1}{r-1}\right)^2 + R^2 \cdot Q^4 \left(\frac{r+1}{r-1}\right)^2\right] \tag{4.79}$$

In the above equations, Q is the Q factor (more precisely, in (4.78) it is the Q factor corresponding to the target BER) and r is the signal extinction ratio. Equations (4.78) and (4.79) do not apply only to RBS, but to any situation where interferometric crosstalk is generated, for example in optical devices like splitters, circulators or OADMs, when a fraction of the signal at one port interferes, due to bad port isolation, with the signal at another port. Another effect that can be modeled as interferometric crosstalk is the FWM.

4.8 Summary

This chapter provided the reader with basic tools to estimate the transmission performance in an optical transport network, and the impact of various propagation effects, enabling the designer to plan an optical network by means of simple analyti-

Table 4.9 Propagation Effects in Optical Fiber

Name	*Acronym*	*Effect*	*Where/When It is More Effective*	*Induced Penalty*
Attenuation	—	Decrease of energy of the propagating signal, due to light scattering or absorption.	In any type of fiber; it is stronger in O band.	Receiver sensitivity penalty.
Chromatic dispersion	CD	Different spectral components of the signal travel at different speed, causing a distortion of the time waveform.	C-band wavelength channels propagating over SMF.	Path penalty, due to receiver sensitivity penalty and ISI.
Polarization mode dispersion	PMD	The two main signal polarization states travel at different speed, causing a distortion of the time waveform	Long-haul systems over old fiber cables.	Q factor penalty, due to receiver sensitivity penalty and ISI.
Self-phase modulation	SPM	Due to the nonlinear nature of the fiber, intensity signal modulation induces a proportional phase modulation, broadening the signal spectrum. In presence of CD, this distorts the time waveform.	C-band wavelength channels transmitted with high launch power over SMF; long-haul links with many amplified spans.	Q factor penalty, due to ISI.
Cross-phase modulation	XPM	Similar to SPM. In DWDM systems, intensity modulation of one channel induces phase modulation in other channels. In presence of CD, this causes crosstalk.	C band wavelength channels transmitted with high launch power over NZ-DSF.	Q factor penalty, due to inter-channel crosstalk.
Four-wave mixing	FWM	Due to the nonlinear nature of the fiber, three optical frequencies generate a fourth frequency, as intermodulation product.	DWDM systems over DSF	Q-factor or OSNR penalty, due to inter-channel crosstalk.
Stimulated Raman scattering	SRS	In DWDM systems, lower wavelength channels transfer power to higher wavelength channels.	DWDM systems, over any kind of fiber.	Sensitivity penalty, due to channel power tilt; Q factor penalty, due to inter-channel crosstalk.
Stimulated Brillouin scattering	SBS	Backscattered high intensity light transfers power from the signal to noise side bands.	High intensity, non-carrier suppressed optical channels.	No significant penalty, applying dithering or suppressing the carrier.
Rayleigh backscattering	RBS	In bidirectional systems over single fiber, backscattered light interferes with signals propagating in its same direction.	Bidirectional fiber systems using the same wavelength in both directions.	Q-factor penalty, due to interferometric crosstalk.

cal models, implementable in a spreadsheet. The considered linear and nonlinear effects, with a qualitative explanation of their impact, are summarized in Table 4.9.

The next chapter will explore the optical technologies and the different network scenarios to which the modeling techniques discussed in this chapter apply.

References

[1] "Optical System Design and Engineering Considerations," ITU-T Supplement G.Sup.39.

[2] "Spectral Grids for WDM Applications: DWDM Frequency Grid," Recommendation ITU-T G.694.1.

[3] Agrawal, G. P., *Fiber-Optic Communication Systems*, MA: John Wiley & Sons, 2002, pp. 23–74.

[4] "Interfaces for the Optical Transport Network," Recommendation ITU-T. G.709/Y.1331.

[5] Ren, C., and A. R. MacKenzie, "Closed Form Approximations to the Error and Complementary Error Functions and Their Applications in Atmospheric Science," *Atmospheric Science Letters,* August 2007.

[6] Marcuse, D., "Calculation of Bit-Error Probability for a Lightwave System with Optical Amplifiers and Post-Detection Gaussian Noise," *Journal of Lightwave Technology,* Vol 9, No. 4, April 1991.

[7] Gaibazzi, A., P. Ghiggino, and N. Shamir, "Performance Calculations of Optically Preamplified IMDD Receivers," *European Transactions on Telecommunicazions,* 1993.

[8] "Avalanche Photodiodes (APDs)–A User Guide," application note from Excelitas Technologies, http://www.excelitas.com/downloads/app_avalanchephotodiodesusersguide.pdf.

[9] Agrawal, G. P., P. J. Anthony, and T. M. Shen, "Dispersion Penalty for 1.3-μm Lightwave Systems with Multimode Semiconductor Lasers," *Journal of Lightwave Technology,* Vol. 6, No. 5, May 1998.

[10] Forestieri, E., "Evaluating the Error Probability in Lightwave Systems with Chromatic Dispersion, Arbitrary Pulse Shape and Pre- and Postdetection Filtering," *Journal of Lightwave Technology,* Vol. 18, No. 11, November 2000.

[11] Kikuchi, N., and S. Sasaki, "Analytical Evaluation Technique of Self-Phase Modulation Effect on the Performance of Cascaded Optical Amplifier Systems," *Journal of Lightwave Technology,* Vol. 13, No. 5, May 1995.

[12] Foschini, G. J., and C. D. Poole, "Statistical Theory of Polarization Dispersion in Single Mode Fibers," *Journal of Lightwave Technology,* Vol. 9. No. 11, November 1991.

[13] Forestieri, E.and , G. Prati, "Exact Analytical Evaluation of Second-Order PMD Impact on the Outage Probability for a Compensated System," *Journal of Lightwave Technology,* Vol. 22, No. 4, April 2004.

[14] Morkel P. R., V. Syngal, D. J. Butler, et al., "PMD-Induced BER Penalties in Optically-Amplified IM/DD Lightwave Systems," *Electronics Letters,* Vol. 30, No. 10, May 1994.

[15] Forestieri E., Prati G., "Novel optical line codes tolerant to fiber chromatic dispersion," Journal of Lightwave Technology, Vol. 19, Issue 11, November 2001.

[16] Ono, T., Y. Yano, and K. Fukuchi, "Characteristics of Optical Duobinary Signals in Terabit/s Capacity, High Spectral Efficiency WDM systems, *Journal of Lightwave Technology,* Vol. 16, No. 5, May 1998.

[17] Smith, N. J., N. J. Doran, W. Forysiak, et al., "Soliton Transmission Using Periodic Dispersion Compensation," *Journal of Lightwave Technology,* Vol. 15, No. 10, October 1997.

[18] Wen, S., and T. Lin, "Ultralong Lightwave Systems with Incomplete Dispersion Compensation," *Journal of Lightwave Technology,* Vol. 19, No. 4, April 2001.

[19] Hui, R., K. R. Demarest, and C.T. Allen, "Cross-Phase Modulation in Multispan WDM Optical Fiber Systems," *Journal of Lightwave Technology,* Vol. 17, No. 6, June 1999.

[20] Inoue, K., and H. Toba, "Fiber Four-Wave Mixing in Multi-Amplifier Systems with Nonuniform Chromatic Dispersion," *Journal of Lightwave Technology,* Vol. 13, No. 1, January 1995.

[21] Eiselt, M., "Limits on DWDM Systems Due to Four-Wave Mixing: A Statistical Approach," *Journal of Lightwave Technology,* Vol. 17, No. 11, November 1999.

[22] Vanholsbeek, F., S. Coen, and P. Emplit, "Raman-Induced Power Tilt in Arbitrarily Large Wavelength-Division-Multiplexed Systems," *Photonics Technology Letters,* Vol. 17, No. 1, January 2005.

[23] Forghieri, F., R. W. Tkach, and A. R. Chraplyvy, "Effect of Modulation Statistics on Raman Crosstalk in WDM Systems," *Photonics Technology Letters,* Vol. 7, No. 1, January 1995.

[24] Aoki, Y., K. Tajima, and I. Mito, "Input Power Limits of Single-Mode Optical Fibers Due to Stimulated Brillouin Scattering in Optical Communication Systems," *Journal of Lightwave Technology,* Vol. 6, No. 5, May 1998.

[25] Pinilla, C. A., "Investigation of Reflective Optical Network Units for Bidirectional Passive Optical Access Networks," philosophy doctor thesis at Universitat Politècnica De Catalunya, June 2007.

CHAPTER 5

Optical Systems and Technologies for Digital Radio Access Networks

5.1 Introduction

This chapter discusses the optical technologies used in three possible deployment scenarios of a RAN; point-to-point fiber systems, DWDM fiber systems and mobile transport over fixed access infrastructure. The choice of the type of system depends on several factors such as presence of preexisting fiber infrastructure, cost needed to upgrade it, total cost of ownership (TCO) of the network, number and geographical density of base stations or radio units, and others. The optical systems described in this chapter fit three different, but equally important, examples of deployment of a RAN: D-RAN with DUs an RUs connected by short haul fiber links (point-to-point fiber systems); C-RAN where the distance between CU and RRUs, and aggregate capacity, can reach several tens of kilometers and several hundreds of Gbit/s, respectively (DWDM systems); and a RAN that reuses installed infrastructure (fiber, central office facilities, etc.) from a preexisting fixed access network, to save the costs of a new installation.

There is no single technology that apply to all those scenarios [1] and the different boundary conditions lead to different design trade-offs. For example, when a fixed access network is already in place, it may be convenient for an operator to reuse it for mobile transport, to avoid new expensive installations. However, bandwidth and latency of a fixed access network could not be adequate for URLLC and eMBB services. The introduction of new technologies is easier in greenfield scenarios, where there are no constraints from legacy networks. For example, optical transmission modules operating at high bit rates, equal to or higher than 25 Gbit/s, are easy to introduce in greenfield DWDM networks but may be not suitable for upgrading the capacity of an existing PON, due to issues in realizing burst mode receivers at high bit rate. Besides latency and capacity, energy efficiency, equipment reliability, network resilience, capability to work in harsh environmental conditions, are all aspects that affect the choice of a networking technology. An important feature in a transport network is the flexibility in configuring bandwidth and connectivity resources. In mobile transport, this allows supporting multiple functional splits, reducing resource overprovisioning and avoiding capacity bottlenecks.

All these aspects will be discussed in this chapter with focus on digital transmission systems. Analog RoF systems, where radio signals are carried over an optical link in their native format, will be discussed in Chapter 7. Point-to-point fiber systems will be described first, illustrating the characteristics of commercial transmitter and receiver modules and some modulation formats of common use. Then, DWDM systems will be addressed. The design of a DWDM system is much more complex than the design of a point-to-point fiber link, since, in addition to transmitters and receivers, DWDM networks make use of devices like optical amplifiers and optical add-drop multiplexers to increase link capacity and distance, respectively. Optical add drop multiplexers, and related technologies, will be described in Chapter 6, while the optical amplifiers will be described in the section of this chapter dedicated to DWDM systems, although their use is possible also in point-to-point fiber links, but they are less common for cost reasons. Finally, the design issues of mobile transport networks over a fixed access infrastructure will be explained, taking it as an example of dimensioning of a backhaul network.

5.2 Point-to-Point Fiber Systems

Apparently, a dedicated fiber link is the most immediate way to connect equipment hosted at two remote sites, such as a DU and an RU; an optical fiber has low attenuation, no practical capacity limit, and is immune to electromagnetic interference and bad weather conditions. However, cabling fiber is expensive, and the return of the investment takes long time. Thus, point-to-point fiber connections are a practical option only when they are facilitated by favorable operational conditions, such as presence of installed unused fibers (dark fibers), presence of ducts that can be filled up with new fibers without initiating new expensive digging and road works, possibility to deploy aerial fiber cables on the power grid, or availability of public funds for new installations. Point-to-point fiber links are common in network scenarios where the connection distance is short and the number of connected units is low, as D-RAN, where no function is centralized and each DU is connected to a small number of RUs in its proximity.

5.2.1 Optical Modules for Point-to-Point Links

Many standards developed by the IEEE 802.3 Ethernet working goup [2] for Local Area Network (LAN) optical interfaces apply also to point-to-point connections in a RAN. In addition, Multisource agreements (MSAs) among manufacturers are in place for the full specification of vendor interoperable optical modules. An MSA takes care of any module design aspect, including physical form factor, optical and electrical interface, optical signal reach, and power consumption. The large datacenter market, and the demand of service providers to develop variants tailored on their specific use cases, encouraged a fragmentation of the optical modules market, leading to many types of optical transceivers, different in transmission wavelength, modulation format, achievable distance, mechanical characteristics, and so on. It is not worthwhile to examine all these variants in detail, considering that this is a market in continuous evolution. Figure 5.1 provides an outlook of the current main options for single mode transmission on optical fiber in O-band (1260 to 1360 nm),

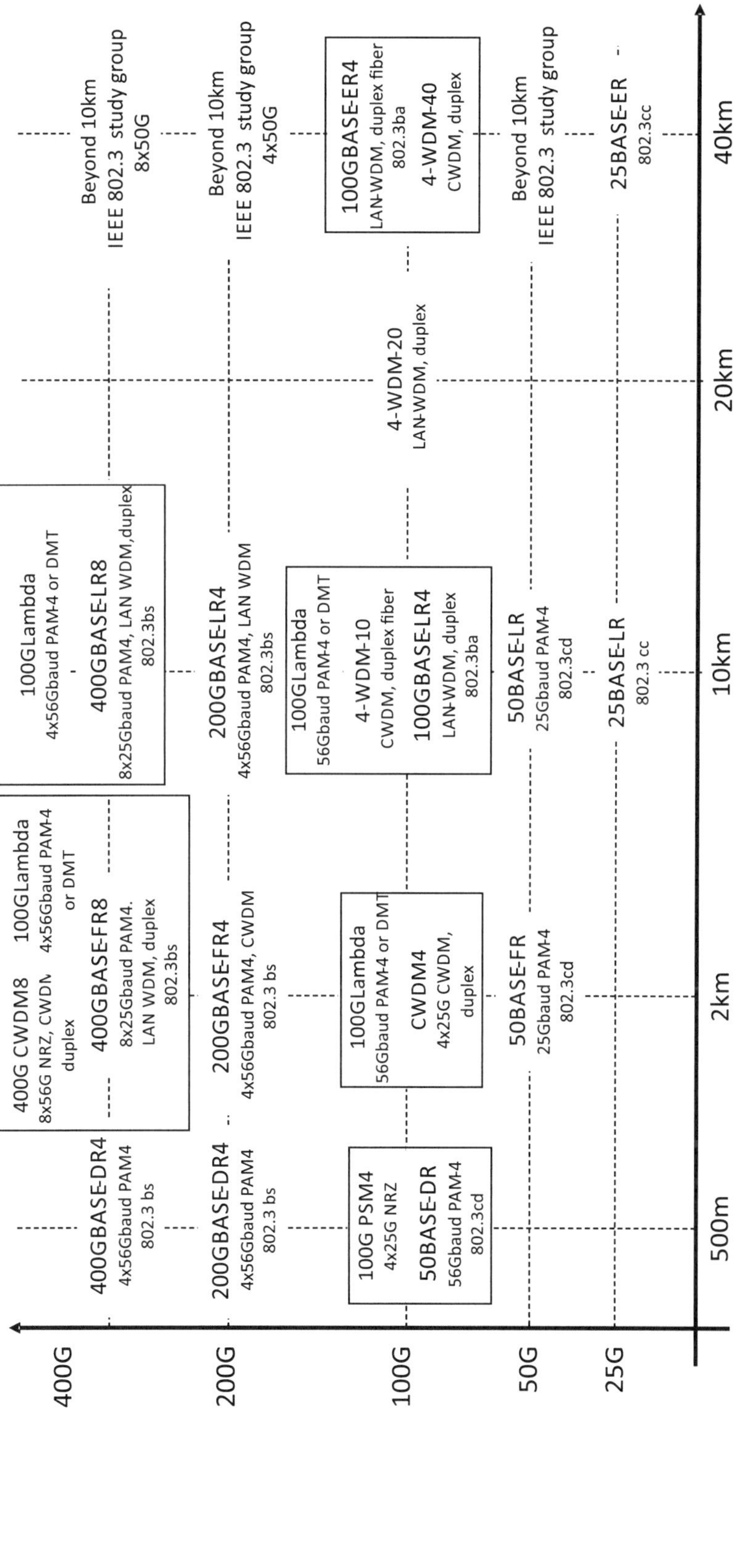

Figure 5.1 high bit-rate optical transceivers for point-to-point fiber links.

at bit rates higher than 10 Gbit/s. Operating in O-band ensures a longer distance, due to the absence of chromatic dispersion.

It is important to highlight some commonalities among the various options in Figure 5.1 [2–7]. All are based on intensity modulation and direct detection. Coherent transmission [8], widely used in long haul optical networks, is not cost effective in a RAN. A cost reduction of the optical coherent modules cannot be excluded in future, but it is today prevented by several factors: Use of I/Q optical modulators to map information bits on both amplitude and phase of the transmitted optical signal; necessity to contain the frequency offset between transmission laser and receiver (LO); use of dual polarization receivers to manage the polarization mismatch between transmission laser and LO; and use of power consuming (tens of Watt) DSP for equalization, polarization recovery, and frequency offset compensation. For similar cost and energy saving reasons, all the options in Figure 5.1 transmit over a single polarization, even if transmitting over two polarizations would allow to double the bit rate without any increase of transmitter and receiver bandwidth. Capacity increase is achieved instead by one of the following techniques, or a combination of them: Coarse wavelength division multiplexing (CWDM); multilevel modulation, such as Order-4 Pulse Amplitude Modulation (PAM-4) or discrete multitone (DMT); and bidirectional transmission over a single fiber (duplex transmission). Strictly speaking, the term CWDM should refer to the frequency grid defined in [9]. Here, it is generically used to indicate wavelength multiplexed lasers, emitting at coarsely spaced optical frequencies, so that no thermal control is needed for frequency stabilization. Examples of CWDM wavelength plans are the LAN WDM grid and the CWDM-4 grid, in Table 5.1.

Usually, an optical transceiver is implemented in a pluggable module format. A pluggable module is a device that can be manually inserted into, and extracted from, the equipment front plane. This simplifies its installation, especially in scenarios where it is difficult to operate, like on the top of an antenna tower. Moreover, it is easy to scale up the capacity of the system, just plugging in new modules, according to a pay-as-you-grow philosophy. Several pluggable modules are specified by MSAs, with the purpose of encouraging an open market based on multiple vendors and interoperable components. A nonexhaustive list is provided in Table 5.2.

5.2.2 Modulation Formats in Point-to-Point Links

In an optical transmission system, the choice of the modulation format is dictated by the trade-off between the achievable performance and the bandwidth limitations introduced by the electronics at transmitter and receiver. If compatible with the current electronic technology, binary modulation formats, such as OOK and

Table 5.1 LAN WDM and CWDM-4 Frequency Grids

Grid	λ_1 Range	λ_2 Range	λ_3 Range	λ_4 Range
LAN WDM	1294.53–1296.59 nm	1299.02–1301.09 nm	1303.54–1305.63 nm	1308.09–1310.19 nm
CWDM-4	1264.5–1277.5 nm	1284.5–1297.5 nm,	1304.5–1317.5 nm	1324.5–1337.5 nm

Table 5.2 Pluggable Optical Modules

Acronym	*Name*	*Bit-Rate (Gbit/s)*	*Typical Power Consumption (Watt)*	*Length × Width × Height Size (mm)*
SFP	Small form factor pluggable	1–2.5	0.5–1	56.5×13.4×8.5
SFP+	Small form factor pluggable	10	1–2	56.5×13.4×8.5
XFP	10 Gbit/ssmall form factor pluggable	10	1.5–3.5	78×18.35×8.5
QSFP	Quad small form factor pluggable	4×2.5	1–2.5	72.4×18.35×8.5
QSFP+	Quad small form factor pluggable	4×10	1.5–3.5	72.4×18.35×8.5
QSFP28	Quad small form factor pluggable	4×28	2.5–4.5	72.4×18.35×8.5
QSFP-DD	QSFP double density	8×28 8×56	12W	72.4×18.35×8.5
CFP	Centum form factor pluggable	4×28 10×10	8–32	144.75×75×13.6
CFP2	Centum form factor pluggable, ½ size	4×28 10×10	3–15	107.5×41.5×12.4
CFP4	Centum form factor pluggable, 1/4 size	4×10 4×28 10×10	1.5–6	88×21.7×9.5

DPSK, are preferred, due to their good sensitivity and implementation simplicity. Optical transmission at bit rates lower than, or equal to, 25 Gbit/s relies indeed on NRZ OOK modulation, which is the simplest modulation format to implement. Due to bandwidth limitations, other techniques are used for transmissions beyond 25 Gbit/s, even if recent technology advances make NRZ viable also at 50 Gbit/s: Higher order modulation, where more than one information bit is carried by each transmitted symbol; parallel transmission over multiple fibers, also called spatial multiplexing; wavelength division multiplexing; and dual polarization transmission. This paragraph focuses on higher order modulation formats suitable of direct detection. They reduce the transmission bandwidth coding a string of bits in a single modulation symbol. This makes them more sensitive to noise and propagation penalties, requiring DSP for equalization and FEC.

Multilevel intensity modulation, indicated as Pulse Amplitude Modulation (PAM), is an example of higher order modulation. Transmitting on M intensity levels, PAM increases the spectral efficiency of a factor $\log_2(M)$ compared to NRZ, for which M=2. However, compared to an NRZ signal with the same average transmitted power, PAM has the disadvantage of decreasing the separation between the signal levels, so impairing the receiver sensitivity. PAM-4 with four amplitude levels is the modulation format usually considered in optical transmission as the right trade-off between performance and spectral efficiency. Compared to NRZ, it doubles the spectral efficiency up to 2 bit/s/Hz, instead of 1 bit/s/Hz, but incurs more than 3-dB sensitivity penalty [10]. The PAM-4 transmitter, shown at the top of Figure 5.2(a), includes the following functional blocks: FEC encoder; bits-into-symbol mapper, based on Gray coding; bandwidth limiting filter, usually based on a FIR structure; DAC; modulator driver; optical modulator; and laser. An advantage of PAM-4 is that the DAC has no stringent requirements. Ideally, one acquired sample per transmitted symbol with a resolution, or effective number of bits (ENOB), equal to two, is sufficient to generate the PAM-4 signal. However, the

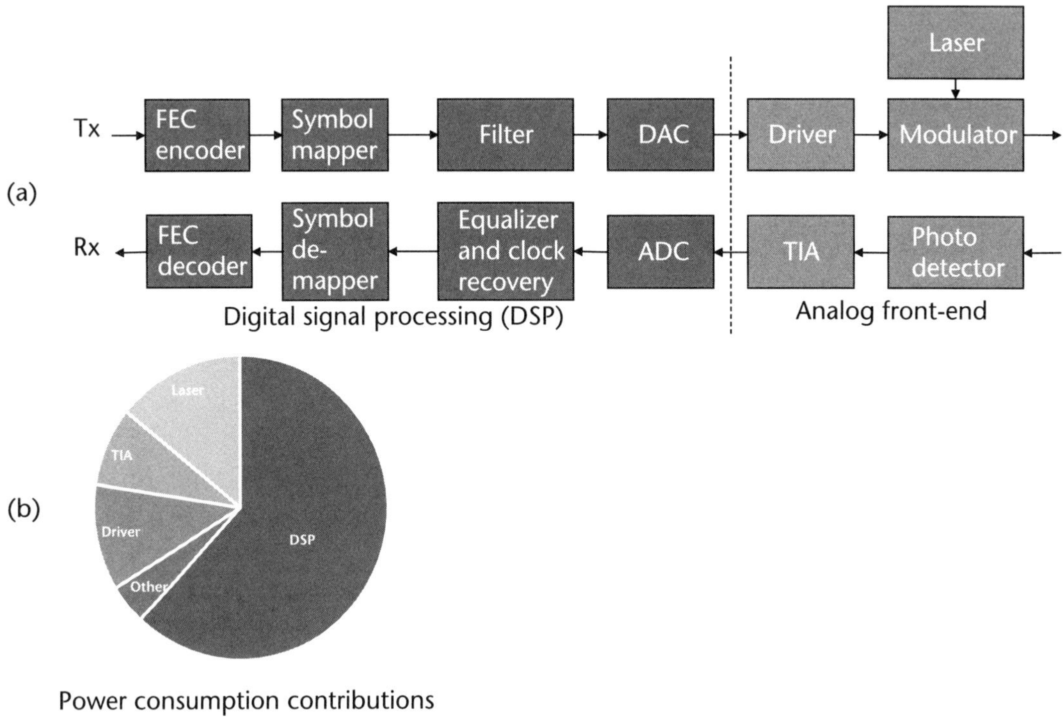

Figure 5.2 PAM-4 (a) transmitter and receiver functional blocks and (b) relative power consumption contributions.

design of the transmitter is sensitive to the linearity of DAC, driver and modulator, which have not to distort the transmitted signal. The PAM-4 receiver at the bottom of Figure 5.2(a) consists of a photodetector, followed by: a transimpedance amplifier (TIA); an ADC; an equalizer, including clock recovery; a symbol-into-bits demapper; and an FEC decoder.

PAM-4 transmitter and receiver are compatible with a QSF28 format, which allows a power consumption of 4.5W. The relative power consumption values of the different blocks are indicatively provided in Figure 5.2(b). In a PAM-4 transmitter, the DAC can be used to optimize the receiver performance depending upon the noise source which is dominating. With thermal noise, which equally affects all signal levels, the best performance is obtained with equally spaced photocurrent levels. Considering that the photocurrent is proportional to the instantaneous power, or intensity, of the optical signal in fiber, this suggests launching in fiber signal levels that are unequally spaced in amplitude but equally spaced in intensity. With ASE noise, which is proportional to the amplitude of the signal, the best performance is obtained when the higher photocurrent levels are spaced more widely than the lower ones. This suggests launching in fiber signal levels that are equally spaced in amplitude, leading to unequally spaced intensity levels. The eye diagrams after the photodetection in the two situations are shown in Figure 5.3.

DMT is a second example of a modulation format suitable of direct detection and, as PAM-4, it is feasible in a QSFP module. DMT is a variant of OFDM, and as OFDM, is based on the transmission of digitally synthetized QAM modulated subcarriers. In a DMT signal, however, power and size of the QAM constellation are individually adjusted on each subcarrier, depending on the measured SNR or BER

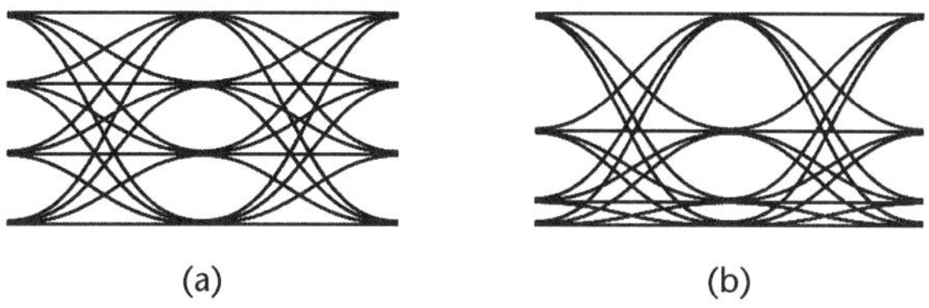

Figure 5.3 Optimal PAM-4 eye diagrams after photodetection in (a) thermal noise dominated regime and (b) ASE noise dominated regime.

(Figure 5.4): Lower-order QAM constellations are used for low SNR subcarriers, whereas higher-order QAM modulation is used for high SNR subcarriers. The size of the subcarrier constellation is calculated at the system start up, during a negotiation phase between transmitter and receiver. The transmitter starts the negotiation, sending a known bit sequence over the maximum size constellation allowed by the system (e.g., 256–QAM); the receiver detects the signal, measures its BER or SNR, and provides a feedback to the transmitter; according to the feedback, the transmitter decides if decreasing or keeping the current constellation size. The process is iterated until the BER or SNR is equal or better than a predetermined threshold. If the threshold is never passed, no data is transmitted over that subcarrier. This happens, for example, when the subcarrier frequency coincides with a notch introduced by the CD on the receiver response [11].

The functional schemes of DMT transmitter and receiver are reported in Figure 5.5.

At the transmitter, input bits are serial-to-parallel (S/P) converted and mapped onto modulation symbols, individually for each subcarrier. Training symbols are added for synchronization and channel estimation purposes. Then, an IFFT is performed and a CP is added, as in OFDM. Finally, the data is serialized in a parallel-to-serial (P/S) conversion block, digital to analog converted in the DAC and sent

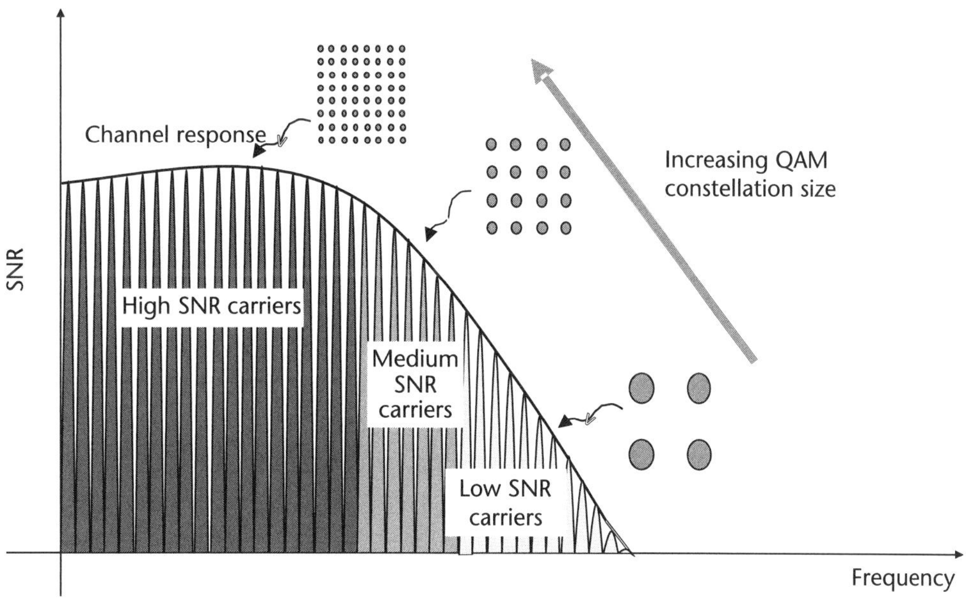

Figure 5.4 Discrete multitone working principle.

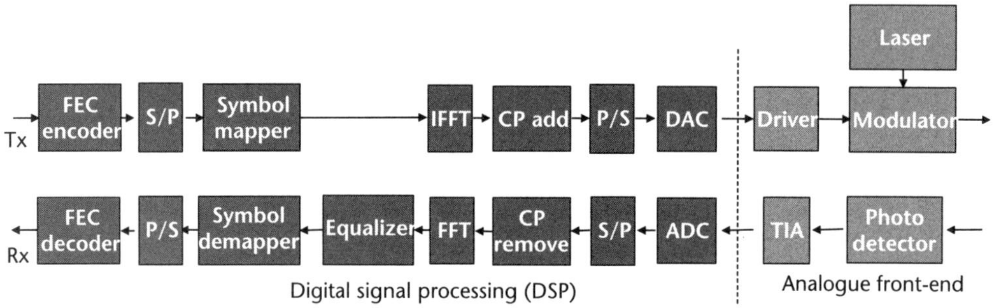

Figure 5.5 DMT transmitter and receiver.

to the analogue transmitter front-end, consisting of driver, modulator and laser. At the receiver, the inverse sequence of operations is performed. DMT is a highly flexible modulation format. It allows adjustment of the bit rate, number of subcarriers and modulation format on each subcarrier, depending on the actual link condition. Moreover, using a small modulation index, a cost-effective implementation is possible, transmitting the DMT signal by means of an intensity optical modulator or a directly modulated laser, and receiving it by means of a single photodiode. Another advantage of DMT is its spectral efficiency, which can be as high as 5 bit/s/Hz [12]. This makes possible the use of low-speed electronics, even at high bit rates. However, as OFDM, DMT suffers from high PAPR values, which leads to a considerable sensitivity penalty: [10] reports that, in thermal noise limited regime, a 100-Gbit/s DMT signal, uniformly loaded with 16-QAM subcarriers, incurs more than a 5-dB sensitivity penalty compared to a 100-Gbit/s NRZ signal. Moreover, the high PAPR requires high resolution DACs and ADCs, with an ENOB of at least 5 or 6 bits.

5.3 Dense WDM Systems

In a DWDM system, multiple modulated optical carriers at different frequencies (or, equivalently, wavelengths) are sent together into the same fiber, to multiply the link capacity. The frequencies of the optical carriers are stabilized—for example, by means of thermoelectric coolers put on the lasers—to prevent any drift that may lead the channels spectra to overlap, causing interference. The commercial DWDM systems operate in the C-band and use the frequency grid defined in [13], with a spacing of 100 or 50 GHz. They have the capability to carry up to 96 channels, 50-GHz spaced. Each channel transmits at a bit rate of 100 Gbit/s, using DP-QPSK modulation with coherent detection, resulting in a total capacity of 9.6 Tbit/s. The use of EDFAs, and the capability of the coherent receiver to compensate for large amounts of accumulated CD, enables the transmission over a few thousand of kilometers. No other communication technology can achieve a comparable performance, neither in capacity nor in distance. The structure of a DWDM link is shown in Figure 5.6.

Optical filters, for example arrayed waveguide gratings (AWG) [14], having a typical insertion loss from 4–6 dB, are used for channel multiplexing and demultiplexing at transmitter and receiver, respectively. Channels can be dropped and added also at intermediate network nodes by means of OADMs. There are two

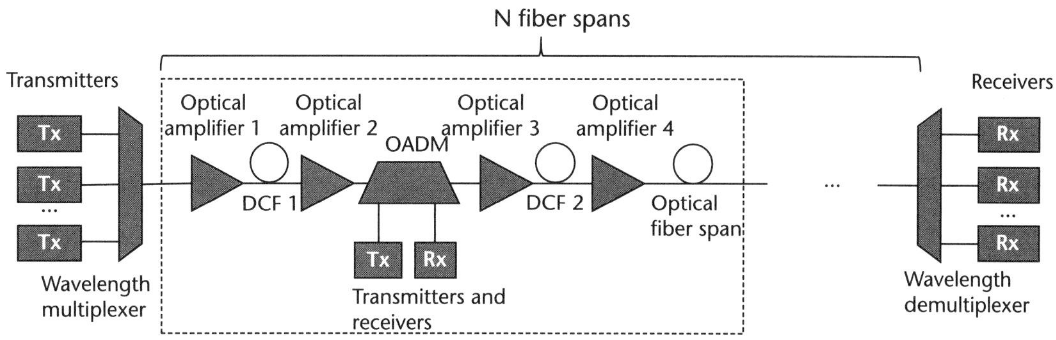

Figure 5.6 DWDM links with optical amplifiers, OADMs, and DCFs.

types of OADM: Fixed OADM (FOADM) and reconfigurable OADM (ROADM). In an FOADM, the number and wavelengths of the added and dropped channels are predetermined. Typically, an FOADM has limited add-drop capability, from 1 up to 8 channels, and is realized with thin-film optical filters or grating filters [15]. The insertion loss of an FOADM is different for added channels, dropped channels, and channels that pass through the FOADM without being added or dropped. The values of insertion loss depend on the technology, the internal structure and the number of channels of the FOADM: typical values are below 4 dB for a number of channels lower than 8. A ROADM is a complex device that can add or drop an arbitrary set of channels, up to the maximum channel count. The set of added and dropped channels can be configured via software, for wavelength rerouting or protection purposes. Several technologies exist to realize a ROADM; an overview is given in [16]. The ROADM enables extremely flexible network planning, where individual wavelength channels can be routed according to the current traffic distribution in the network. Today, due to cost issues, the use of ROADMs is mostly confined in WANs, but cost-effective integrated photonic implementations have been proposed to target access and aggregation networks, including the RAN [17]. However, other cost reduction actions accepting some decrease in performance, are necessary to extend the use of the DWDM technology to the RAN. For example, in a RAN, 100 GHz is more suitable as frequency spacing than 50 GHz, to avoid high accuracy but expensive laser stabilization. For similar cost-saving reasons, direct detection optical modules in SFP format (Table 5.2), are preferred to coherent optical modules. Direct detection modulation formats suitable for a DWDM RAN will be discussed in Section 5.3.4. First, optical amplifiers are described in Section 5.3.1, a statistical design approach to DWDM links is illustrated in Section 5.3.2 and the issues of wavelength dependent losses and gains in DWM systems are discussed in Section 5.3.3.

5.3.1 Optical Amplifiers

Optical amplification is achieved through different physical effects, such as Population inversion in doped fibers or waveguides, obtained by means of optical pumping; population inversion in semiconductor structures, obtained through electrical pumping; FWM in fiber or other nonlinear materials; and Raman or Brillouin scattering in fiber. The erbium-doped fiber amplifier (EDFA) [18] is the most used type of optical amplifier in optical networks. It finds large use in DWDM networks, thanks to the wide amplification bandwidth (about 30 nm), the low noise figure (typically 6 dB), the high gain (>30 dB), and the high output power (>20 dBm). The typical response time of an EDFA is in the order of 10 ms. This makes the EDFA gain and output power insensitive to fast variations of the input power, as those due to the channel modulation. Consequently, in an amplified DWDM system, the modulation data pattern on one channel leads to no undesired intensity modulation, or crosstalk, onto the other channels. Despite these nice features, the optical amplifiers are preferably avoided in a RAN, for cost-saving reasons. The cost reduction resulting in other optical components from the application of integrated photonic technologies does not apply to the EDFA, which is basically a coil of erbium-doped fiber coupled with a pump source, typically a high-power laser emitting at 980 or 1480 nm. Furthermore, the EDFA only works in C-band, where the necessity to use dispersion compensation modules leads to additional costs.

Operation in the O-band and integrated photonic implementation are both possible instead with semiconductor optical amplifiers (SOAs). An SOA is a device consisting of several semiconductor layers, designed to confine the electrons in an active region, where they are excited to obtain light amplification by means of stimulated emission. In an SOA, the electrons are pumped onto the excited level by injecting electric current. Basically, an SOA can be considered as a laser without reflective facets. However, a SOA has a fast, subnanosecond response time, which makes it unsuitable for DWDM systems, where it introduces significant cross-gain modulation penalty [19]. In principle, the use of SOAs in a DWDM system would still be possible operating in small signal gain regime, that is at low input powers, where the SOA gain is insensitive to fluctuations of the input power. However, low values of input power lead to low values of OSNR, prejudicing the system feasibility in many practical cases. For these reasons, the typical use of an SOA is at the transmitter or the receiver of a single channel, to increase the transmitter output power or to improve the receiver sensitivity, respectively, and with the warning that putting an SOA at the receiver could lead to the introduction of significant amounts of noise at low received powers, which may limit the achievable sensitivity gain with noise sensitive modulation formats like PAM-4 or DMT.

Raman fiber amplifiers [20] are a third type of amplifier used in optical networks. Due to the low noise figure, they find application in unrepeated submarine links or in long haul multispan links, typically as the first stage of hybrid Raman/EDFA line amplifiers. Raman amplification requires expensive, high-power unpolarized pump lasers, an aspect which prevented, so far, their application on a large scale.

A mathematical model, applicable to any type of optical amplifier, will be illustrated in the following to calculate the optical power of signal and noise in an optical communication system with multiple amplified fiber spans. The model is largely used to estimate the performance of a DWDM system.

An optical link, consisting of a Tx and an Rx, separated by M optically amplified fiber spans, can be modeled as shown in Figure 5.7. The spans may include passive devices, like attenuators, OADMs and DCF. The attenuation of the kth span ($k=1...M$) is indicated with A_k, while G_k and NF_k are the gain and noise figure of the kth optical amplifier. The signal power at the output of the kth amplifier is P_k, and P_0 is the transmitter output power.

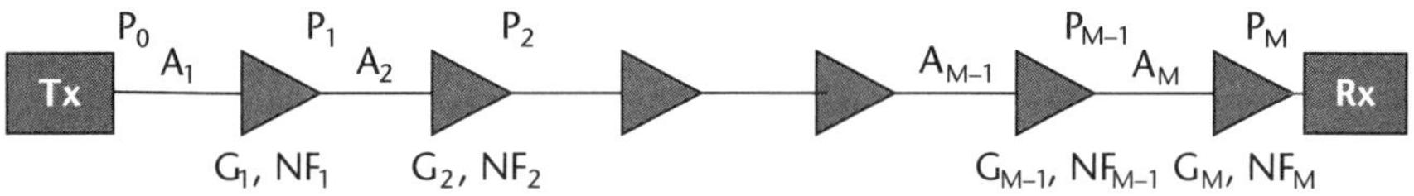

Figure 5.7 Optically amplified system.

It can be demonstrated, using quantum mechanics fundamental principles [21], that the psd of the ASE noise generated by each optical amplifier is

$$N_0 = h \cdot f \cdot NF \cdot (G-1) \tag{5.1}$$

where h=6.62607004·10^{-34} Joule/s is Planck's constant, f is the optical signal frequency, NF is the amplifier noise figure and G is the amplifier total gain, defined as ratio between total input and output power, including both signal and noise. Due to the ASE noise contribution to the amplifier output power, the signal gain G_S is lower than the total gain G. Indicating with P_{in} and P_{out} the total input power and the total output power of the amplifier, total gain and signal gain are given by

$$G = \frac{P_{out}}{P_{in}} \quad G_S = \frac{P_{out}}{P_{in} + N_0 \cdot B_{eq}} \tag{5.2}$$

where B_{eq} is the equivalent noise bandwidth of the amplifier and N_0 is the ASE psd.

In the amplified link of Figure 5.7, the output power of each amplifier is the sum of three terms: Signal output power; ASE noise generated by amplifiers preceding the current optical amplifier in the chain and amplified by it; and ASE noise generated by the amplifier itself. Making the realistic assumptions that the signal power is much greater than the noise power, so that G~G_S, and that the amplifier gain is high, so that G>>1, the OSNR at the receiver can be calculated as follows:

$$OSNR = P_0 \frac{\prod_{m=1}^{M} G_m / A_m}{\sum_{m=1}^{M} N_m \cdot \prod_{k=m+1}^{N} G_k / A_k} \tag{5.3}$$

In (5.3), N_m= N_{0m}·B, where N_{0m} is the ASE psd of the mth amplifier and B is the OSNR resolution bandwidth. In the following, (5.3) will be applied to the three notable cases of a single preamplified span, a single span with a booster amplifier, and a multispan link with all equal spans.

Preamplifier is the term used to identify an optical amplifier in front of a receiver. The OSNR of a preamplified single span link (Figure 5.8) can be calculated putting $M = 1$ in (5.3).

$$OSNR = P_0 \frac{G_1 / A_1}{N_1} = P_0 \frac{G_1 / A_1}{k \cdot NF_1 \cdot G_1} = \frac{P_0}{k \cdot NF_1 \cdot A_1} \tag{5.4}$$

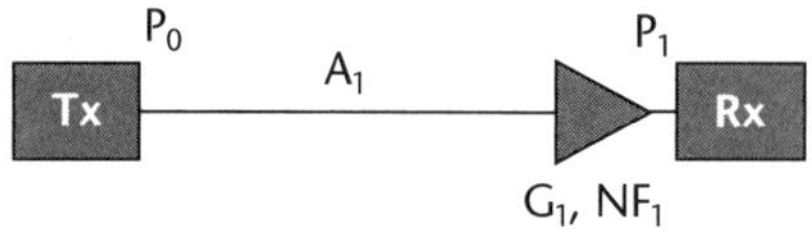

Figure 5.8 Preamplified link.

Where k = h·f·B. If B = 12.5 GHz and f = 193.6 THz, k = 1.58 nanowatt, or –58 dBm. Hence, in the C-band, the OSNR of a preamplified link can be expressed in decibels as:

$$OSNR_{dB} = P_0 + 58 - NF_{dB} - A \tag{5.5}$$

For example, the OSNR is 32 dB for a link with A = 20 dB, P_0 = 0 dBm, and NF = 6 dB (a typical value for an EDFA amplifier). Note that in a preamplified link, the OSNR does not depend on the amplifier gain.

If the optical amplifier is placed after the transmitter rather than before the receiver, it is called a booster amplifier (BA). The structure of a link with a BA is shown in Figure 5.9, and the OSNR is calculated in (5.6).

$$OSNR = P_0 \frac{G_1 / A_1}{N_1 / A_1} = \frac{P_0}{kNF_1} \tag{5.6}$$

Comparing (5.6) and (5.4), it can be noted that the OSNR is higher with a BA than with a PA. On the other hand, the received power is lower with a BA of the same factor, that is, the link attenuation.

In an optical link with multiple spans, it is a common design practice to set all the amplifiers at the same output power in such a way that the amplifier gain exactly compensates for the preceding span loss and G_m/A_m=1 in (5.3). If all the amplifiers and spans in the link are equal, (5.3) simplifies as follows:

$$OSNR = \frac{P_0}{kNFAM} = \frac{P_0}{kNFe^{\alpha L}M} \tag{5.7}$$

where α is the span attenuation coefficient and L is the span length. Equation (5.7) shows that the OSNR decreases proportionally to the number of spans M, but exponentially with the span length. Consequently, an optical communication system with many short spans performs better than a system with fewer longer spans. This is evident from Figure 5.10, which shows the span attenuation and the span length necessary to obtain an OSNR of 20 dB for multiple span links with NF = 6 dB, P_0 =

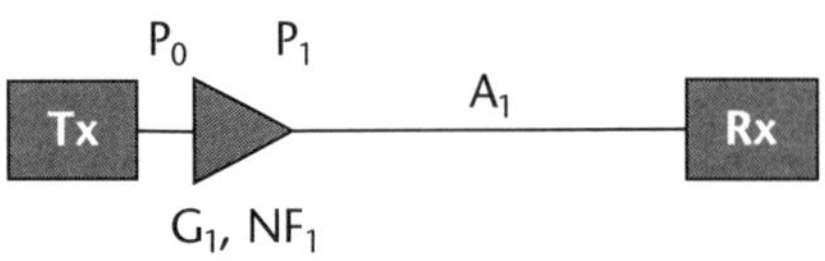

Figure 5.9 Link with a booster amplifier.

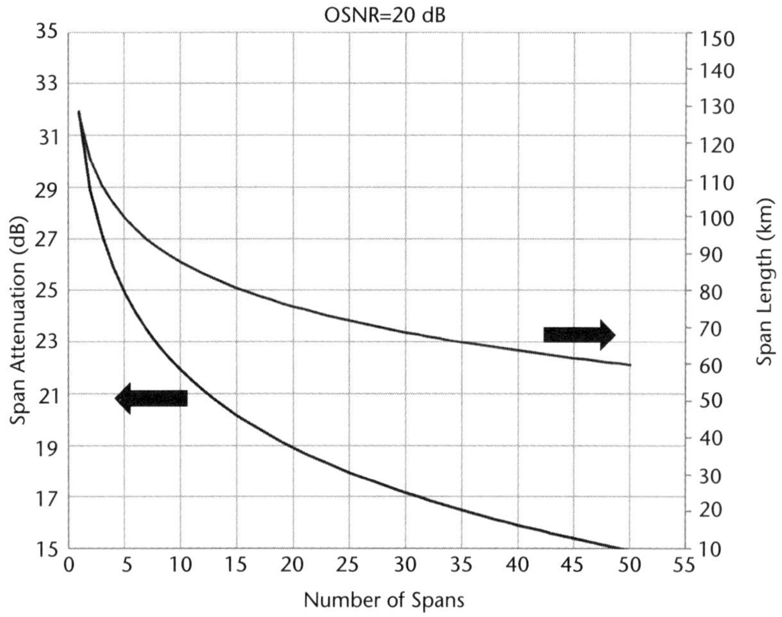

Figure 5.10 Span length and attenuation to obtain an OSNR of 20 dB.

0 dBm, and $\alpha = 0.25$ dB/km. With a single span, the target OSNR is achieved with 127.7 km and 31.9 dB of span length and span attenuation, respectively. The same OSNR is attained by increasing the number of spans to 50 and decreasing their length and attenuation to 59.7 km and 14.9 dB, respectively—achieving a total length of 2985 km.

5.3.2 Statistical Design of DWDM Links

A worst-case approach is usually adopted in the design of passive point-to-point optical links. For example, to calculate the minimum received power, the maximum link attenuation is subtracted from the minimum transmitted power. Similarly, the maximum received power is obtained by subtracting the minimum link attenuation from the maximum transmitted power. However, in optical links consisting of many cascaded elements, a worst-case design leads to overestimate and underestimate, respectively, the maximum and minimum received power, and to state the unfeasibility of links that would perfectly work in practice. A typical example of optical link where a worst-case design does not apply is a DWDM link consisting of a wavelength multiplexer, multiple fiber segments, OADMs, optical amplifiers, DCFs, and wavelength demultiplexer (Figure 5.6). The design of such a link involves long summations to calculate received optical power, OSNR, accumulated chromatic dispersion and polarization mode dispersion. Calculating the received power of a channel added at one of the OADMs in the link as example, it is possible to write the following equation, where all losses and gains are in decibels.

$$Rx\,Power = Tx\,Power - L_{add} - \sum_{\substack{number\ of\\ fiber\ spans}} L_{fiber\ span} + \sum_{\substack{number\ of\\ amplifiers}} G_{amplifier} - \sum_{\substack{number\ of\\ DCFs}} L_{DCF} - \sum_{\substack{number\ of\\ OADMs}} L_{pass-through} - L_{drop} \quad (5.8)$$

where L_{add} is the loss of the OADM at which the channel is added (or the wavelength multiplexer loss); $L_{fiber\ span}$, L_{DCF}, and $L_{pass-through}$ are, respectively, the attenuations of fiber spans, DCFs, and OADMs included between the OADM port where the channel is added and the drop port of the OADM at which it is dropped; $G_{amplifier}$ are the gains of the optical amplifiers included between the add port and the drop port; and L_{drop} is the loss of the OADM at which the channel is dropped (or the wavelength demultiplexer loss). Assuming $N = 10$ in Figure 5.6, a channel traveling from the extreme left to the extreme right of the link would experience the attenuation of 10 fiber spans, 10 OADMs, and 20 DCFs, and the gain of 40 optical amplifiers. A worst-case design is clearly unsuitable in this situation. A statistical design is therefore used, with all losses and gains considered as independent random variables. The mean values of the variables are summed up (the gains are considered as negative losses), as well as their variances (variances are positive for both gains and losses). Thus the mean value η and variance σ^2 of the total link loss are obtained, and the minimum value and the maximum value of received power can be estimated as:

$$\begin{aligned} min\,Rx\,Power &= min\,Tx\,Power - \eta - n \cdot \sigma \\ max\,Rx\,Power &= max\,Tx\,Power - \eta + n \cdot \sigma \end{aligned} \quad (5.9)$$

In (5.9), the multiplier of the standard deviation n is chosen according to the confidence level the designer is willing to accept. Assuming a Gaussian distribution (an assumption justified by the central limit approximation for a link consisting of many cascaded elements), $n = 3$ corresponds to a confidence level of 99.7%. Unfortunately, the designer seldom knows the statistical distribution of the loss of the fiber and the components in the link, for various reasons, such as the component manufacturer only providing minimum and maximum loss values, or the data sheet of the fiber is no longer available, and so forth. In these cases, mean value and variance need to be estimated by means of heuristics. For example, if the minimum and maximum loss, L_{min} and L_{max} respectively, are available for a component, the following approximations can be made to calculate average value and standard deviation of its insertion loss:

$$\begin{aligned} \eta &\sim \frac{L_{min} + L_{max}}{2} \\ n \cdot \sigma &\sim \frac{L_{max} - L_{min}}{2} \end{aligned} \quad (5.10)$$

A similar statistical analysis, which will not be repeated here, can be performed for the accumulated chromatic dispersion in fiber spans and DCFs.

5.3.3 Wavelength Dependent Losses and Gains

The insertion loss of a passive device, the gain of an optical amplifier and the fiber attenuation coefficient are functions of the wavelength, and hence they are different for the different channels of the DWDM spectrum. For a passive device, such as a FOADM or a multiplexer, the manufacturer usually specifies the uniformity, defined as the maximum difference of insertion loss between an arbitrary pair of wavelengths of the DWDM spectrum. Typical values of uniformity are included between 0.5 and 2 dB. The curve of the fiber attenuation coefficient versus the wavelength is qualitatively reported in Figure 4.3 in Chapter 4. In C-band, it can be approximated with a straight line and specified by giving two points. For example, in the fiber specified in [22], the estimated difference between the attenuation coefficients at 1,525 and 1,575 nm is 0.02 dB/km.

SRS also introduces a tilt of the channel powers, which depends on the total input power in fiber. For equally spaced DWDM channels, the tilt can be calculated according to (5.11) [20]:

$$\left(\frac{P_N}{P_1}\right)_{dB} = 2.17 \cdot \left(\frac{g'}{A_{eff}}\right) \cdot P_0 \cdot N \cdot (N-1) \cdot \Delta f \tag{5.11}$$

where $g'/A_{eff} = 6.4 \cdot 10^{-17}$ m^{-1}W^{-1}Hz^{-1}; P_1 and P_N are the input powers of the lowest frequency channel and highest frequency channel of the DWDM spectrum; N is the number of channels; and Δf is the channel frequency spacing.

The gain spectrum of an optical amplifier shows a strong dependence on the amplifier total gain. In an EDFA operating with a high pump power (higher total gain), shorter wavelengths around 1,530 nm are amplified more than longer wavelengths, around 1,560 nm. Decreasing the pump power (and the total gain) the gain slope is inverted. Mathematical models have been developed to explain this behavior, but they are very sensitive to parameters, such as the concentration of the erbium ions in the doped fiber, which are not known with sufficient accuracy. Therefore, black-box models [23, 24] are used more frequently. A black-box model uses a couple of reference gain spectra, measured in known working conditions, and infers by interpolation the gain at a different operating point. An example of a black-box model is the tilt function model reported in (5.12). It uses two gain spectra, $G_{1,dB}(\lambda)$ and $G_{2,dB}(\lambda)$, measured at two different total gains, to estimate the tilt function $T(\lambda)$, which can be interpreted as the slope of the gain variation versus the total gain. Another gain spectrum $G_{ref,dB}(\lambda)$ is used as baseline, together with a reference wavelength, λ_{ref}, inside the DWDM spectrum.

$$\begin{gathered} G_{dB}(\lambda) = T(\lambda) \cdot \left[G_{dB}(\lambda_{ref}) - G_{ref,dB}(\lambda_{ref})\right] + G_{ref,dB}(\lambda) \\ T(\lambda) = \frac{G_{1,db}(\lambda) - G_{2,db}(\lambda)}{G_{1,db}(\lambda_{ref}) - G_{2,db}(\lambda_{ref})} \end{gathered} \tag{5.12}$$

An advantage of the black-box models is that they also work for optical amplifiers having a complex internal structure, with several gain stages and optical attenuators, used to adjust the amplifier gain.

In an amplified DWDM link, any difference of optical power among wavelengths (sometimes referred as gain flatness or gain variation) leads to a corresponding variation of OSNR. It is observed in real systems that both optical power difference and OSNR difference, in decibels, increase linearly with the number of spans, but the OSNR difference grows, approximately, at half the rate of the optical power difference. This behaviour can be intuitively explained considering a link consisting of N equal spans, having an attenuation equal to a, in linear units. Each span is preceded by an optical amplifier, having a total gain g, which exactly compensates the span attenuation ($g = a$). However, due to the gain flatness, different wavelengths experience different gains, generally different from g. Two specific wavelengths are considered in the following: A first wavelength, having gain g_1 lower than g, and a second wavelength whose gain g_2 obeys the equation $g_1 \cdot g_2 \sim g$. Obviously, $g_2 > g$. This working assumption is equivalent to assume that the gain spectrum is symmetric around the average gain g. Defining $x = g_2/g = g/g_1$, after N span, the power ratio between the two wavelengths will be proportional to x^{2N}. Since the ASE noise at a certain wavelength is proportional to the gain at that wavelength, summing up all noise contributions after N amplified spans, the accumulated noise will be proportional to $x + x^2 \ldots + x^N$, for the second wavelength, and to $x^{-1} + x^{-2} \ldots + x^{-N}$, for the first wavelength. Applying the properties of the geometric series, it easy to demonstrate that the ratio between the two summations is x^{N+1}. Hence, the OSNR ratio between the two wavelengths will be proportional to $x^{2N}/x^{N+1} = x^{N-1}$. Moving from liner units to decibels, and defining ΔP and $\Delta OSNR$ as the difference of optical power and OSNR for the two channels after N spans, it is

$$\Delta P \propto 2 \cdot \mathrm{N} \quad \Delta OSNR \propto (N-1) \tag{5.13}$$

which is what we intended to demonstrate.

Preemphasis techniques are commonly applied to equalize the OSNR of DWDM channels. A simple iterative method is reported in [25]. At each iteration step, k, the transmitted channel powers P_1, P_2, ..., P_n ...P_N are updated according to (5.14):

$$P_n^{(k)} = P_0 \frac{P_n^{(k-1)} / OSNR_n^{(k-1)}}{\sum_{n=1}^{\mathrm{N}} \frac{P_n^{(k-1)}}{OSNR_n^{(k-1)}}} \tag{5.14}$$

where P_0 is the nominal transmitted power, fixed by design, and $P_n^{(k)}$ and $OSNR_n^{(k-1)}$ are optical power and OSNR of the nth channel, at the steps k and (k-1) respectively. The OSNR preemphasis also mitigates the channel power difference [26]. More sophisticated preemphasis techniques [27, 28] have been developed to deal with situations where the OSNR is not the dominant effects but other factors, such as nonlinear propagation in fiber, significantly contribute to the BER.

5.3.4 Modulation Formats in a DWDM RAN

Direct detection modulation formats, like PAM-4 and DMT (Sections 5.2.2), originally developed for single channel links in O-band, also apply to DWDM systems using a laser emitting in C-band. However, two additional impairments, not present in O-band, must be considered: CD and, in amplified optical links, amplification noise.

Figure 5.11 shows the receiver sensitivity penalty versus the fiber length, at a bit rate of 50 Gbit/s and a reference BER of 10^{-3} for direct detection modulation formats suitable to be used in a RAN. All penalty values are referred to the OOK back to back sensitivity, so that the OOK penalty at 0 km is 0 dB. Fiber chromatic dispersion coefficient and attenuation coefficient are 17.5 ps/(nm·km) and 0.22 dB/km, respectively.

Due to its wide spectrum, OOK with NRZ pulse coding shows a sudden increase of penalty even with short fiber lengths and, to be used, it requires CD compensation, which is problematic in a RAN, since placing DCMs of different size at each network node oblige the network operator to keep their inventory and store spare parts. Tunable dispersion compensation modules [29] overcome this issue but introduce further active devices in the network. Moreover, both fixed and tunable DCM in line cannot exactly compensate the dispersion for channels that are added and dropped at different nodes and experience different propagation paths. Electronic [30] and optical [31] tunable dispersion compensation devices embedded in the transceiver are preferable under all the above aspects. An integrated optical component based on silicon nitride microrings is reported in [32].

The other binary format shown in Figure 5.11 is DBPSK, or DPSK. It shows a marginal improvement compared to OOK, with slightly better back-to-back penalty and CD tolerance.

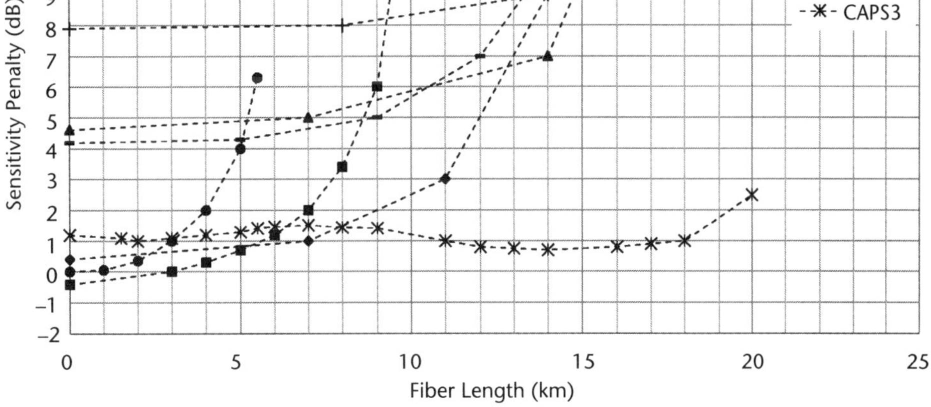

Figure 5.11 Optical penalty for various direct detection modulation formats.

PAM-4 starts from 4 dB of back-to-back penalty compared to OOK, due to the smaller distance between the signal amplitude levels, but it recovers the gap within the first 5 km of fiber, due to its better spectral efficiency.

DQPSK shows a marginal improvement compared to PAM-4 for distances longer than 11 km, which is probably not enough to justify its use in a mobile transport network, due to its implementation complexity. It requires an I/Q modulator at the transmitter, and interferometers and balanced photodiodes at the receiver.

The latest two modulation formats reported in Figure 5.11 are Order-1 and Order-3 Combined Amplitude Phase Shift, CAPS1 and CAPS3, respectively. They belong to a class of line codes, of which duobinary is a special case [33], designed to counteract the ISI induced by CD. As for optical duobinary, a simple implementation exists for CAPS1: Both can be generated by narrow filtering of a binary signal (OOK for duobinary and DPSK for CAPS1). CAPS1 has a negligible back-to-back penalty compared to OOK and extends its reach from 4 up to 10 km, with 2 dB of dispersion penalty. CAPS3 requires instead an I/Q modulator to be generated [34], exhibiting a penalty smaller than 2 dB over the first 10 km of fiber.

DMT is not reported in Figure 5.11, since its performance depends on implementation variables such as number of subcarriers, QAM constellation size, number of training symbols, cyclic prefix length, and others. Reference [35] reports 40 km achieved with a 49.6 Gbit/s DMT signal.

OSNR is the other issue that affects the performance of DWDM systems with direct detection. Figure 5.12 reports measured BER versus OSNR for OOK, PAM-4 and CAPS3. The OSNR is referred to the conventional 0.1-nm resolution bandwidth. The received power (P_{RX}) is indicated for all the curves. The parameters α and β, which are the signal weights on the I/Q modulation arms [34], are also reported for CAPS3.

The measurements in Figure 5.12 are performed in back-to-back conditions, but CAPS3 is also measured by inserting 10 km of fiber, to verify its robustness to

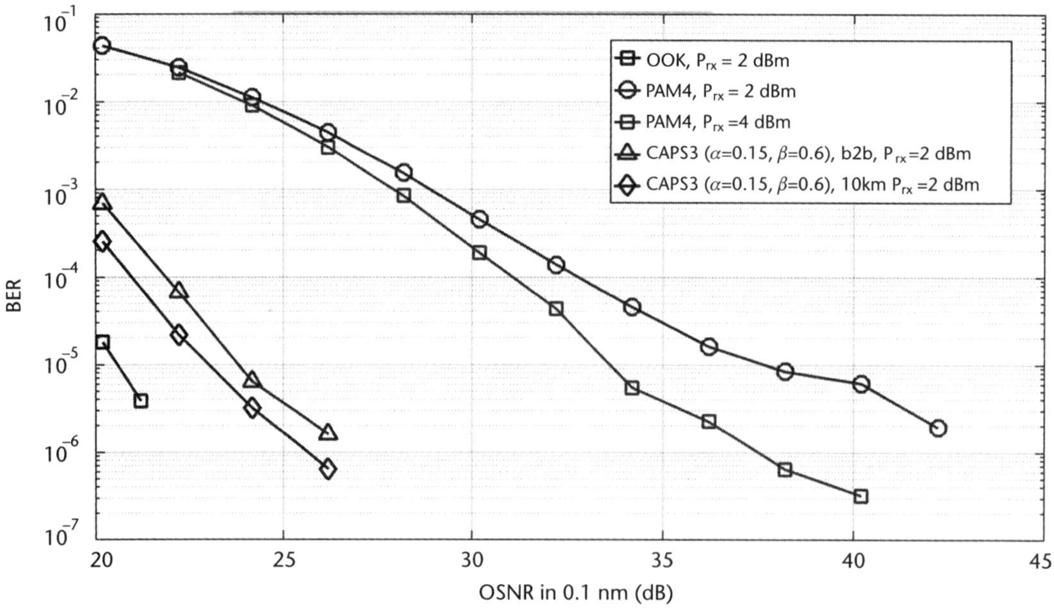

Figure 5.12 BER versus OSNR for direct detection modulation formats.

CD. At 10^{-5} BER, the PAM-4 penalty is about 12 dB compared to OOK, while the CAPS3 penalty is approximately 2.5 dB in back to back and about 3 dB in presence of fiber.

The DMT performance in presence of OSNR has been analyzed in [12]: As expected, due to the high PAPR, the required OSNR is high, always greater than 35 dB.

5.3.5 Further Considerations on DWDM RANs

Operational and maintenance costs are important aspects to consider when selecting a technology for a RAN. Due to the high number and high geographical density of network nodes, technologies that simplify the installation procedures, reduce the installation times, and minimize the number of spare parts, are highly preferred. One critical aspect of the DWDM systems is the high number of wavelengths, which requires a proportional number of optical modules spare parts and labeling all transmitting and receiving ports in the network with the wavelength value. To solve this issue, the concept of colorless DWDM network, based on wavelength port agnostic devices, was introduced. The term colorless was first applied to DWDM PON systems, to indicate transmitters, based on various technologies [36] and capable of automatically adjusting the emitted wavelength based on the multiplexer port they were connected to. A similar concept has been developed in ITU-T G.698.4 [37], which defines wavelength tuning and power adjustment procedures for a transmitter at the tail-end equipment of a point-to-multipoint DWDM network, not to interfere with adjacent channels. In ITU-T G.698.4, the tuning mechanism is coordinated by hub node, or head-end equipment. The concept of colorless devices can be further extended to include tunable dispersion compensation modules and ROADMs, since both these devices allow to use the same reconfigurable hardware at every network node, simplifying network provisioning and installation. An overview of tunable dispersion compensation techniques has been provided in Section 5.3.4. ROADMs based on microelectromechanical systems (MEMS) or liquid crystal technologies are already used in WANs, but they are too expensive for a RAN. An example of cost-effective ROADM based on silicon photonic system-on-chip technology is reported in [17].

To efficiently exploit the capacity offered by a DWDM system, and lower the cost per transmitted bit/s, the direct mapping of a client signal over a wavelength is suggestible only for high, constant bit rate client signals, or when latency constraints make any further signal processing not applicable. For this reason, wavelength multiplexing is often the lowest layer of a multilayer architecture, where it coexists with upper circuit or packet multiplexing layers. Circuit multiplexing is used for constant bit rate signals having moderate capacity, which are suitable to be aggregated in a higher bit rate wavelength channel by means of TDM. It is also used for transport interfaces that require a deterministic propagation delay through the network, a requirement which might be difficult to satisfy in a packet network because of the presence of random delays due to packet buffering and routing. Different circuit multiplexing standards exist; one, limited to CPRI client signals, is described in [38]. The OTN standard [39] specifies instead a generic multiplexing hierarchy, including payload encapsulation, operation administration and maintenance (OAM) overhead, and FEC. CPRI was added recently to the OTN clients list [40]. Due to the complexity of the OTN multiplexing protocol, and issues in guaranteeing clock frequency accuracy and small differential delay between uplink and downlink, as required by time sensitive fronthaul signals, an evolution of the OTN standard, tailored on 5G transport applications, is under study at ITU-T. The Flex Ethernet[41] standard, developed by the Optical Internetworking Forum (OIF) and natively conceived to efficiently multiplex Ethernet framed client signals, is another example of a circuit multiplexing scheme considered for mobile transport networks. Finally, packet switching allows for fully exploiting the statistical multiplexing gain, transmitting at a bit rate proportional to the current data load. Packet networks are especially suitable for RAN interfaces, which are not latency sensitive when the split point if outside the HARQ loop (see Chapter 3), or in distributed network scenarios characterized by low or moderate traffic load so that the probability that the packet switches experience congestion is low. Though packet networks were not conceived to deal with time-sensitive client signals, standards to provide time deterministic services in Ethernet networks, guaranteeing packet transport with bounded latency, low packet delay variation, and low packet loss, were developed by the IEEE Time-Sensitive Networking (TSN) Task Group. The Deterministic Networking (DetNet) working group at the Internet Engineering Task Force (IETF) has a similar task for IP-based and multiprotocol label switching (MPLS) networks.

5.4 Mobile Transport over Fixed-Access Networks

5.4.1 Passive Optical Networks

A fixed access network can be based on point-to-point fiber links or on one of the PON standards developed by IEEE and ITU-T. The technologies for point-to-point fiber links were discussed previously in this chapter, so this section will focus on mobile transport over PON, taking as reference the ITU-T standard. PON topology and working principle are illustrated in Figure 5.13.

A PON consists of a hub node, called an optical line terminal (OLT), located at the service provider central office (CO); a number of optical network units (ONUs) located at the user premises; and an optical distribution node (ODN). The ODN includes a trunk fiber, a 1:N optical power splitter, and N drop fibers, where N is the maximum number of connected ONUs. In the downstream (DS) direction, from the OLT to the ONUs, the transmission is based on a TDM protocol: Different time slots in a single broadcasted frame are assigned to different ONUs. Each ONU processes its slot and discards the other ones. In the opposite upstream (US) direction, a time division multiplexing access (TDMA) approach is used, whereby the OLT assigns to each ONU a time window it can transmit data within. No transmission is allowed to the ONU out of the assigned window so that, if there are no data to transmit, no signal is received at the OLT. Moreover, due to the different length of the drop fibers, data bursts received at the OLT, coming from different ONUs, have different optical power, as shown in Figure 5.13. This makes necessary the use of burst mode receivers [42] at the OLT, which are able to deal with sudden variations of input powers. The burst mode receiver is the most critical device of a PON and it is the major obstacle to scale its operation to bit rates higher than 10

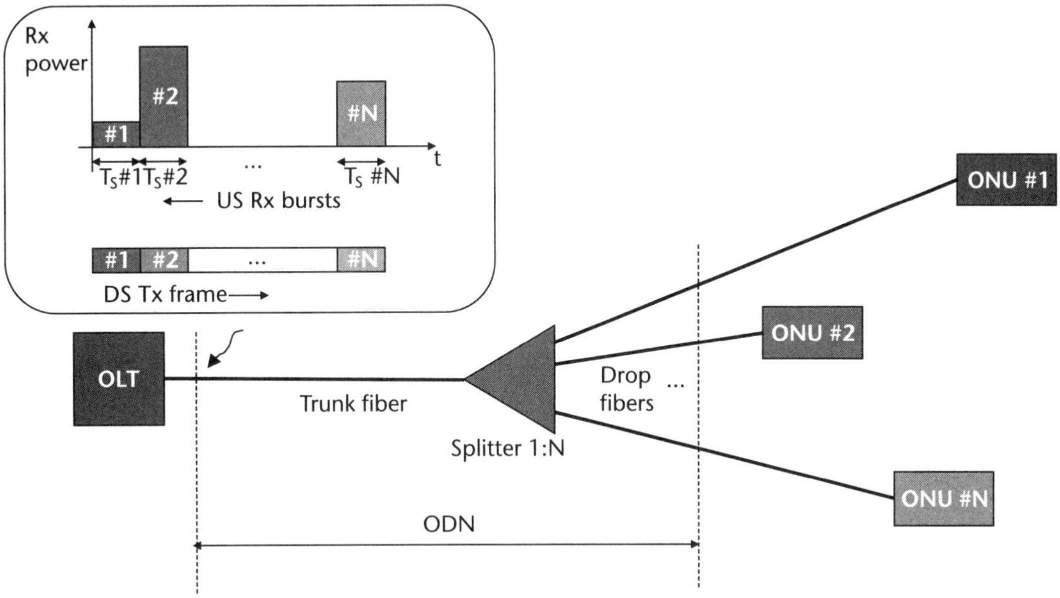

Figure 5.13 PON working principle.

Gbit/s. As mentioned, the OLT is responsible for assigning transmission windows to the ONUs. To avoid collisions, the OLT measures the upstream transmission delay from each ONU to the OLT, in such a way that the ONUs have a common time reference. The time alignment procedure is called ranging. Once the upstream delays are known, the OLT gives the permission (grant) to the ONU to transmit in a certain time interval. To efficiently exploit the upstream bandwidth, the grants are reassigned every few milliseconds. The class of algorithms used by the OLT to assign bandwidth to the ONUs is known as dynamic bandwidth assignment (DBA). A DBA partitions the upstream bandwidth among the different ONUs, based on the actual traffic load and transmission demands. At this purpose, the DBA uses the transmission containers (T-CONTs), which are logical connections, whose instances are first created by each ONU during its activation, and then discovered by the OLT. One ONU can generate several T-CONTs, having different priority or traffic class. Each T-CONT corresponds to a specific upstream transmission interval.

The second next-generation PON (NG-PON2) standard adds a WDM layer to the TDM layer for multiplexing traffic from different ONUs (Figure 5.14).

This approach is called time-wavelength division multiplexing (TWDM). In Figure 5.14, two ONUs share the same wavelength, λ_1, by using a TDM/TDMA protocol, as in a regular PON. Similarly, three ONUs share another wavelength, λ_2. Using an optical band split filter connected to a WDM multiplexer (Mux), it is possible to upgrade the capacity of the system, adding further wavelengths for point-to-point connections. In Figure 5.14, one of the ONUs uses one of these wavelengths.

There are several generations of PON [43] and a comprehensive description is outside of the scope of this book. Table 5.3 summarizes the main physical layer specifications of the ITU-T PON standards.

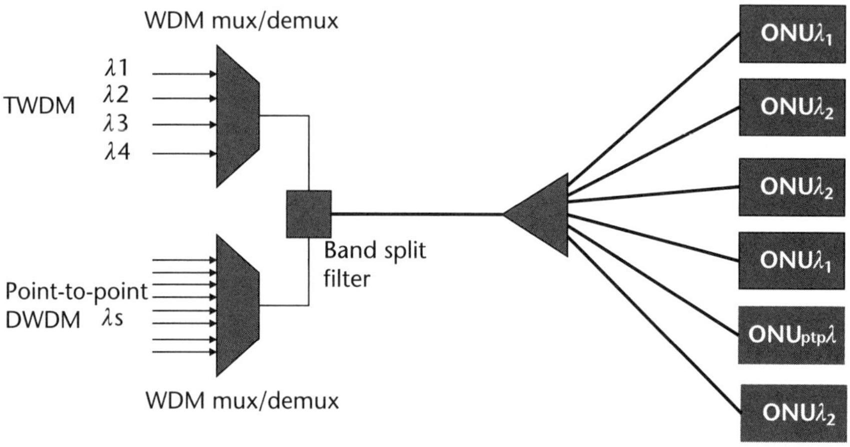

Figure 5.14 NG-PON2 working principle.

Table 5.3 Physical Layer Characteristic of PON Systems

	GPON	*XG-PON*	*XGS-PON*	*NG-PON2*
ITU-T Standard	G.984	G.987	G.9807	G.989
DS/US Bit Tate in Mbit/s	2488.32/2488.32	9.85328/2488.32	9.85328/9.85328	4x9.85328/9.85328
Access Type	TDM	TDM	TDM	TWDM
Split Ratio	1:64	1:64	1:64	1:64
Distance in km	20	40	20	40
Max ODN Loss	30 dB	35 dB	29 dB	35 dB

5.4.2 Mobile Transport over PON

Uses cases and technologies for mobile transport over PON are addressed in [44], for both high- and low-level functional splits. Two macro scenarios are considered by the ITU-T: Fixed access services carried over legacy TDM PON, with WDM overlay reserved to wireless services; and a PON completely dedicated to wireless services. In the former scenario, new wavelengths like point-to-point wavelengths in an NG-PON2 (Figure 5.14) are used for wireless services, without sharing bandwidth with the fixed access services of the legacy PON. In the second scenario, where the PON is deployed only for wireless services, dedicated wavelengths are used for latency demanding interfaces, while services that are compatible with the delay introduced by the DBA (some milliseconds) are carried using a TDM/TDMA protocol. Bandwidth assignment mechanisms are proposed in [44] to improve the upstream latency in a TDM PON, which can be several milliseconds, because of the grant procedures. A first method to reduce the upstream latency of wireless services is to establish service classes, assigning to mobile traffic the highest priority. For example, four service classes may be set: Fixed bandwidth, assured bandwidth, nonassured bandwidth, and best-effort. The fixed bandwidth class has the highest priority and periodically assigns bandwidth regardless of the presence of traffic demands. The assured bandwidth class is similar, but the bandwidth is allocated only on demand. The remaining bandwidth is assigned, on demand, to the latest two classes. A second method to reduce the upstream latency is known as cooperative DBA. In a cooperative DBA, the mobile network uses the PON downstream signal to send allocation requests of upstream bandwidth. The requests are made in advance, so that the mobile traffic experiences no waiting queue at the ONU. The time diagram in Figure 5.16 illustrates the operation of a cooperative DBA, with reference to the simple system of Figure 5.15, where DU and RU are connected to OLT and ONU, respectively. In a more realistic scenario, several RUs are connected to the same ONU, but the fundamental DBA mechanisms do not change.

A cooperative DBA assumes that mobile equipment and PON have a common time reference, so that the time scale can be divided in slots of equal duration that can be recognized by both systems. As shown in Figure 5.16, at a certain time slot, denoted by M, the UE sends a traffic request for a future time slot, M+N. The request travels along the system and arrives to the DU, where it is processed by the mobile scheduler, which decides to send a grant that comes back to the UE through the network. Concurrently, the DU notifies the OLT that a grant has been assigned (hence the name cooperative DBA), so that the OLT can schedule an upstream time

Figure 5.15 Mobile equipment connected to a PON.

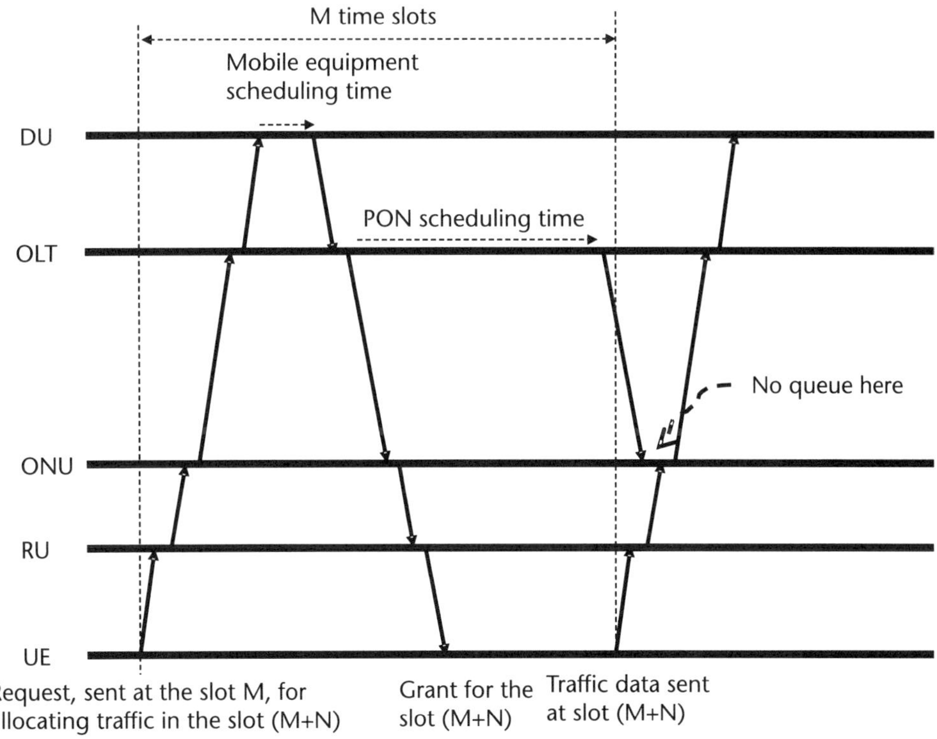

Figure 5.16 Cooperative DBA time diagram.

slot for traffic sent by the ONU at the time (M+N)·T + D, where T is the time slot duration and D is the delay from the UE to the ONU. Thus when the data is sent by the U, they arrive to the ONU and can pass through it, without being queued.

5.4.3 Dimensioning of a Backhaul Network

The backhaul traffic from a base station to the core network has no tight latency requirements, so it is expected that most of the mobile traffic sent over a PON will be backhaul traffic. In this section, a PON will be used as an example for dimensioning a backhaul network with two aggregation stages, but the following considerations are valid in general. The reference network scenario is illustrated in Figure 5.17: Several radio BSs are connected to each ONU, which acts a first aggregation stage. Then, the aggregated traffic is conveyed to the OLT, where a second traffic aggregation is performed before sending it to the core network.

For each cell of the mobile network, peak throughput R_{peak}, mean throughput value μ_{cell}, and throughput standard deviation σ_{cell}, are defined. The mean cell throughput is the average capacity of the cell, inferred by statistical measurements, when many users share the same air interface resources under one sector. The throughput standard deviation is derived similarly. A gaussian distribution is assumed for the cell throughput so that these two parameters are sufficient to define its probability distribution. The cell peak throughput is the air interface capacity between BS and UE, on one sector. It depends on the air interface bandwidth, or more precisely, on the number of resource blocks (see Chapter 1); but there is no immediate relationship between the two variables, because the achievable throughput varies with the channel propagation conditions. To calculate the actual throughput, the mobile terminal estimates and communicates to the BS a number, called channel quality indicator (CQI). Based on the CQI, the base station selects from a lookup table another number, called a modulation and coding

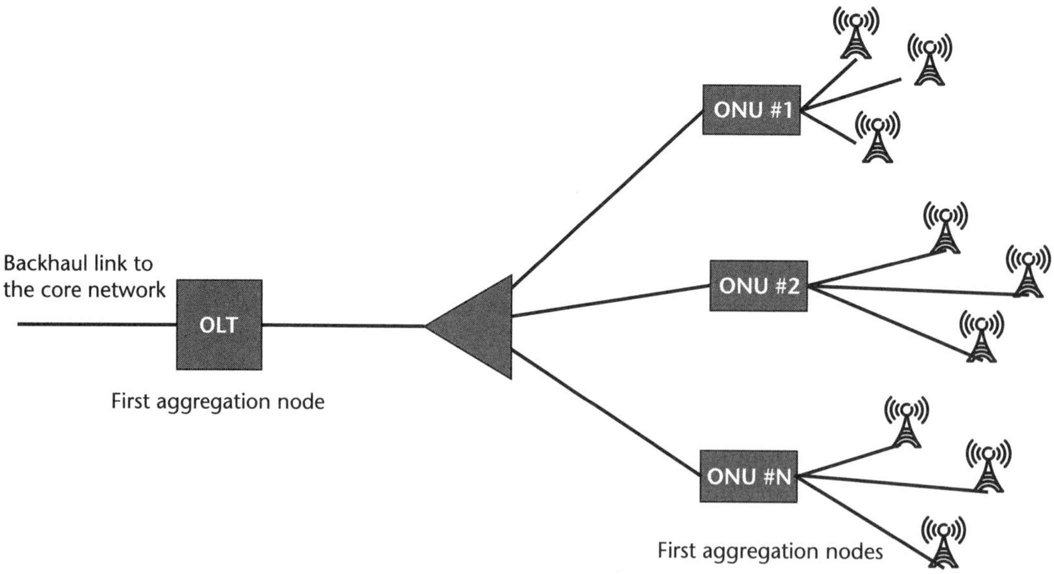

Figure 5.17 PON-based backhaul network.

scheme (MCS), which is used to define how many bits are carried by each transport block. The CQI to MCS mapping rule depends on the design rules of the radio equipment vendor. Two further tables (e.g., Table 7.1.7.1-1 and Table 7.1.7.2.1-1 in [45]) are needed to calculate the throughput. The first table maps the MCS into QAM modulation order and transport block size (TBS) index. The second table maps the TBS index into the number of bits per transport block, that is, the TBS. For example, with MCS = 28, it is TBS Index = 26 and TBS=75376 bits. Hence, in one TTI (1 ms), the throughput of a 20 MHz LTE system without MIMO is 75.376 Mbit/s. The throughput proportionally scales with the air interface bandwidth and the number of MIMO layers.

In a backhaul network, the first aggregation node is typically dimensioned to support the peak rate of the connected base stations. The capacity C_1 of each last-mile link between BS and first aggregation node is then:

$$C_1 = k \cdot \max\left(R_{peak}, n \cdot \mu + \sqrt{n} \cdot f \cdot \sigma_{cell}\right) \tag{5.15}$$

where n is the number of cells connected to each BS, k is a transport overhead factor (k~1.1-1.3) and f is the factor to obtain a desired percentile from the probability distribution curve.

The capacity C_2 of the second aggregation node, as well as that of any further aggregation stage, is not dimensioned on the peak rate but relies on statistical multiplexing assumptions, such as:

$$C_2 = c \cdot k \cdot \left(N_{RBS} n \cdot \mu + \sqrt{n \cdot N_{RBS}} \cdot f \cdot \sigma_{cell}\right) \tag{5.16}$$

where N_{RBS} is the number of connected BSs and c is the statistical multiplexing concentration factor ($0<c<1$, $c = 1$ corresponds to no statistical multiplexing).

5.5 Summary

This chapter provided an overview of the optical transmission technologies for a D-RAN, addressing performance issues as well as other important operational issues. Three relevant examples were provided as deployment scenarios: Point-to-point passive link-based D-RAN, C-RAN based on DWDM technology, and mobile transport over fixed access infrastructure. The concept of multilayer transport was also introduced during the DWDM networks discussion. Enabling optical switching technologies will be illustrated in the next chapter. The case of mobile transport over fixed access infrastructure has been taken as an example of bandwidth assignment and capacity dimensioning of a backhaul network.

References

[1] "D2.1: Study and Assessment of Physical and Link Layer Technologies for Crosshaul," Report of H2020 5G-Crosshaul project, Grant No. 6715982017.

[2] IEEE 802.3 Ethernet Working Group, http://www.ieee802.org/3/.

[3] *100G PSM4 MSA Web page:* http://psm4.org/.

[4] *CWDM8 MSA Web Page:* https://www.cwdm8-msa.org/.

[5] *CWDM4 MSA Web Page*: http://www.cwdm4-msa.org/.

[6] *4WDM MSA Web Page*: http://4wdm-msa.org/.

[7] *100G Lambda MSA Web Page:* http://100glambda.com/.

[8] Kikuchi, K., "Fundamentals of Coherent Optical Communication," *Journal of Lightwave Technology,* Vol. 34, No. 1, January 2016.

[9] "Spectral Grids for WDM Applications: CWDM Wavelength Grid," Recommendation ITU-T G.694.2.

[10] Cole, C., I. Lyubomirsky, A. Ghiasi, et al., "Higher-Order Modulation for Client Optics," *Communications Magazine,* Vol. 51, No. 3, March 2013.

[11] Srinivasan, R.C., and J. C. Cartledge, "On Using Fiber Transfer Functions to Characterize Laser Chirp and Fiber Dispersion," *Photonics Technology Letters,* Vol. 7, No. 11, November 1995.

[12] Nadal, L., M. Svaluto Moreolo, J. M. Fàbrega, et al., "DMT Modulation with Adaptive Loading for High Bit Rate Transmission Over Directly Detected Optical Channels," *Journal of Lightwave Technology,* Vol. 32, No. 21, November 2014.

[13] "Spectral Grids for WDM Applications: DWDM Frequency Grid," Recommendation ITU-T G.694.1.

[14] Venghaus, H. (ed.), Wavelength Filters in Fibre Optics, Berlin: Springer, 2006, pp. 125–187.

[15] Kagawa, M., H. Tsukada, and M. Yoneda, "Optical Add-Drop Multiplexers for Metro/Access Networks," *Furukawa Review,* No. 23, 2003.

[16] Marom, D. M., P. D. Colbourne, A. D'errico, et al., "Survey of Photonic Switching Architectures and Technologies in Support of Spatially and Spectrally Flexible Optical Networking," *Journal of Optical Communication and Networking,* Vol.9, No. 1, January 2017.

[17] Iovanna, P F. ., Cavaliere, and F. Testa, "Future Proof Optical Network Infrastructure for 5G Transport," *Journal of Optical Communications and Networking,* Vol. 8, No. 12, December 2016.

[18] Agrawal, G. P., *Fiber-Optic Communication Systems*, Hoboken, NJ: John Wiley & Sons, 2002, pp. 226–273.

[19] Ramaswami, R., and P. A. Humblet, "Amplifier Induced Crosstalk in Multichannel Optical Amplifier," *Journal of Lightwave Technology,* Vol. 8, No. 12, December 1990.

[20] Bigo, S., S. Gauchard, A. Bertaina, et al., "Experimental Investigation of Stimulated Raman Scattering Limitation in WDM Transmission over Various Types of Fiber Infrastructure," *Photonics Technology Letters,* Vol. 11, No. 6, June 1999.

[21] Kweon, G., "Noise Figure of Optical Amplifiers," *Journal of the Korean Physical Society,* Vol. 41, No. 5, November 2002.

[22] Corning® SMF-28® Ultra Optical Fiber–Product Information, https://www.corning.com/media/worldwide/coc/documents/Fiber/SMF-28%20Ultra.pdf.

[23] Burgmeier, J., A. Cords, and R. Marz, "A Black Box Model of EDFA's Operating in WDM Systems," *Journal of Lightwave Technology,* Vol. 16, No. 7, July 1998.

[24] Lazaro Villa, J. A., F. Bonada Bo, V. P. Qureol, et al., "Extended Black-Box Model for Fiber Length Variation of Erbium-Doped Fiber Amplifiers," *Photonics Technology Letters,* Vol. 20, No. 24, December 2008.

[25] Chraplyvy, A. R., J. A. Nagel, and R. W. Tkach, "Equalization in Amplified WDM Lightwave Transmission Systems," *Photonics Technology Letters,* Vol. 4, No. 8, August 1992.

[26] Cavaliere, F., G. Bruno, and G. Razzetta, "Incremento della distanza di trasmissione nei sistemi WDM tramite tecniche di controllo spettrale," *Proc. Fotonica 2003*, Riva del Garda (Italy), April 2003.

[27] Ciaramella, E., L. Giorgi, A. D'Errico, et al., "A Highly Effective Technique for Setting the Power Preemphasis in WDM Optical Systems," *Journal of Lightwave Technology,* Vol. 24, No. 1, January 2006.

[28] Ciaramella, E., S. Brugioni, E. Matarazzo, et al., "Automated Performance Equalization in WDM Networks with Inline Optical Add Drop Multiplexers," *Proc. National Fiber Optic Engineers Conference*, San Diego, February 2008.

[29] Painchaud, Y., M. Lapointe, F. Trépanier, et al., "Recent Progress on FBG-Based Tunable Dispersion Compensators for 40 Gb/s Applications," *Proc. Optical Fiber Communication Conference 2007*, Anaheim, CA, March 2007.

[30] Foggi, T., E. Forestieri, and F. Colavolpe, "Maximum-Likelihood Sequence Detection with Closed-Form Metrics in OOK Optical Systems Impaired by GVD and PMD," *Journal of Lightwave Technology*, Vol. 24, No. 8, August 2006.

[31] Sorianello ,V., G. De Angelis, F. Fresi, et al., "100Gb/s PolMux-NRZ Transmission at 1550 nm over 30 km Single Mode Fiber Enabled by a Silicon Photonics Optical Dispersion Compensator," *Proc. 2018 Optical Fiber Communications Conference and Exposition (OFC)*, San Diego, March 2018.

[32] Fresi, F., M. Morsy-Osman, F. Forestieri, et al., "50 Gb/s Transmission over Uncompensated Link up to 20 km Exploiting DSP-Free Direct-Detection," *Proc. 2018 Conference on Lasers and Electro-Optics (CLEO)*, San Jose, CA, August 2018.

[33] Forestieri, E., and G. Prati, "Novel Optical Line Codes Tolerant to Fiber Chromatic Dispersion," *Journal of Lightwave Technology*, Vol. 19, No. 11, November 2001.

[34] Forestieri, E., M. Secondini, F. Fresi, et al., "Extending the Reach of Short-Reach Optical Interconnects with DSP-Free Direct Detection," *Journal of Lightwave Technology*, Vol. 35, No. 15, August 2017.

[35] Lacava, C., I. Cardea, I. Demirtzioglou, et al., "49.6 Gb/s Direct Detection DMT Transmission over 40 km Single Mode Fibre Using an Electrically Packaged Silicon Photonic Modulator," *Optics Express*, Vol. 25, No. 24, November 2017.

[36] Ponzini, F., F. Cavaliere, G. Berrettini, et al., "Evolution Scenario toward WDM-PON," *Journal of Optical Communications and Networking*," Vol. 1, No. 4, September 2009.

[37] "Multichannel Bi-Directional DWDM Applications with Port Agnostic Single-Channel Optical Interfaces," Recommendation ITU-T G.698.4.

[38] "CPRI Interface Specification", http://www.cpri.info/spec.html.

[39] "Interfaces for the Optical Transport Network," Recommendation ITU-T. G.709/Y.1331.

[40] "OTN Transport of CPRI Signals," Supplement ITU-T G.Sup.56.

[41] "Flex Ethernet 2.0", OIF Implementation Agreement, June 2018.

[42] Chao, S., C. Lian-Kuan, and C. Kwok-Wai, "Theory of Burst-Mode Receiver and Its Applications in Optical Multiaccess Networks," *Journal of Lightwave Technology*, Vol. 15, No. 4, April 1997.

[43] Nesset, D., "PON Roadmap," *Journal of Optical Communication and Networking*, Vol. 9, No. 1, January 2017.

[44] "5G Wireless Fronthaul Requirements in a Passive Optical Network Context," Supplement ITU-T G.sup.66.

[45] "Evolved Universal Terrestrial Radio Access (E-UTRA)," Physical Layer Procedures," 3GPP Technical Specification 36.213.

CHAPTER 6

Optical Switching for Radio Access and Aggregation Networks

6.1 Introduction

The RAN trend is to offload into optics the large bandwidth demand, while aggregating the different network traffic independent of user location [1]. The optical networks trend is to support a wider variety of functionality evolving to provide an intelligent platform for various types of services delivery [2, 3]. Reconfigurability of next-generation RANs can be efficiently provided at the physical layer with the introduction of efficient and cost-effective optical switches to route broadband data transmission.

This chapter is divided into three main sections. First, Section 6.2 discusses the network applications of optical switches, highlighting issues and benefits in different optical node realization. Then Section 6.3 presents the current state of the art of optical switching elements, and focuses on the new solutions in silicon photonics technology aiming to overcome all the issues in a radio access and aggregation network, proposing a simple example of switching element, the microring resonator, treated in an analytical model. Finally, Section 6.4 analyses the silicon photonic technology approach with application examples: Simple examples of ROADM nodes in silicon photonics are presented and detailed as optimized for different network topologies in the radio access and aggregation network. Further technological and implementation details are reported in Appendixes 6A and 6B.

6.2 Network Application of Optical Switches

The traffic carried by radio access and conveyed to and from the broadband aggregation optical networks, as well as the per-channel interface rates required by routers, are growing at a remarkable pace, typically 30% of compound annual growth rate (CAGR) [1]. Optical transmission and networking advancements have so far satisfied these traffic requirements by delivering data over the transport network infrastructure in a cost- and energy-efficient manner, but the cost remains not sustainable for radio and cloud network applications, and needs to be scaled [1].

Furthermore, we are approaching the fundamental limits of both single-mode fibers and the growing capabilities of conventional networks. Every two years, we create more data than we have created in all of history — therefore, the evolution of current technologies could be not enough in the long-term run.

5G technology demands for optical fibers, which suddenly increases the demand for optical switches for fast routing and low latency and broadband transceivers. While the recent advancements and commercial availability of optical communications can significantly boost the network capacity, a further increase will be very challenging, mostly when performing signal routing, switching, and processing in the overall network [3].

It is worth noting that the bandwidth demands of mobile broadband networks are currently either saturated or are growing slowly; but on the contrary, due to the introduction of high-speed data services (LTE and 5G), the amount of bandwidth to be backhauled for data is increasing rapidly, and due to this increasing demand for new high-speed services in the mobile network, a new technological breakthrough in broadband radio access networks is needed.

Any means to expand the network as the number of subscribers increases and to provide additional services, without requiring an increase in operating expenses, represents a crucial solution that cannot be underestimated. In a recently proposed evolution scenario, reconfigurable optical networks based on WDM are proposed instead of fixed point to point interconnections [3]. To realize reconfigurable optical networks, reconfigurable optical nodes are used. With the 5G network, the availability of fiber becomes fundamental, while the demand for fiber-based backhaul and fronthaul continues to increase with aggregation points in access networks that are distributed in wider areas and should be interconnected with a flexible and reconfigurable topology.

To this aim, WDM proves to be the most suitable architecture, allowing flexible wavelength allocation for fronthaul and backhaul applications. With WDM network architectures, several reconfigurable interconnection solutions are possible; with fiber-to-fiber switching, it is possible to reconfigure the network, but the system data throughput can be scaled up only linearly by augmenting the number of installed fibers, becoming a nonpractical solution for telecom grade applications, especially in the RAN. Conversely, if N wavelengths can be multiplexed over a single optical fiber to scale up the overall system capacity, the number of fibers to be installed can be kept low. In the latter case, the wavelength switching can be the best option to perform network reconfigurability (i.e., with much lower traffic granularity handling with respect to the fiber switch case).

Among the possible network evolutions, a long-term trend may also involve multifiber (fiber overlays) systems with space division multiplexing (SDM) systems based on multicore fibers (MCF), few-mode fibers (FMF) or few-mode multicore fibers (FM-MCF). Today is possible to manufacture fibers supporting multiple spatial modes or containing multiple spatially separated cores, which could be used as parallel channels for independent signals. The SDM systems are based mainly on this potential, and can dramatically increase the overall system capacity to overcome network bottlenecks for broadband applications in the next years to come. The mentioned evolution is still under research evaluation by system vendors and network operators and requires not yet estimated investments in the network infrastructure and still needs assessment in terms of costs and benefits trade-off.

6.2.1 Network Reconfigurability

In Figure 6.1, an example of cell distribution with 5G system deployment is reported, as well as two main network segments directly connected with RAN: Fronthaul and backhaul networks. The illustrated example depicts the importance of the availability of optical fibers and switching nodes, while the demand for fiber-based backhaul and fronthaul continues to increase, with a wide distribution of new aggregation nodes. The backhaul interfaces the fronthaul with a metro network and consists of traffic aggregation nodes (i.e., aggregated traffic is handled for network reconfiguration and routed toward the metro network), typically interconnected in ring topologies and point to point links. The metro network segment is the part of the telecommunication network connecting localized networks with the core network — the highway of traffic transport for the wide delivery of Telecom and Datacom applications.

6.2.2 Optical Nodes

Two main types of optical nodes can be defined in optical networks. One is based on optical to electrical to optical (OEO) signal conversions, and the network reconfiguration is performed with an electrical circuit switch, called the cross-point switch. The other is based on all-optical switches (OOO) with no need to apply the electro-optical-electrical (EOE) conversion to route signals in the optical network. In both cases, a ROADM of signals is performed to select or generate traffic locally. These solutions define two different ROADM network nodes, which should be capable of routing continuously increasing capacity signals in the shortest time possible. The switching time is strictly related to the used technology. Already installed systems in metro and core networks use optical switches with reconfiguration time of the order of few milliseconds (e.g., using micromechanics or liquid crystals on silicon (LCoS) pixels), but future optical switches (e.g., those based on integrated photonics technology) will introduce fast time network reconfigurability thanks to the higher switching speed.

The next optical switches must be suitable for both Datacom networks and Telecom networks, favoring efficient traffic handling and enabling software-defined networking in both domains to optimize network resource utilization and managing of the quality of delivered services in the overall network. The network latency corresponds to the service request and delivery time from and to the final user. The latency can be determined by the end-to-end signal propagation time, the traversed network nodes reconfiguration time, and the data processing time. The shorter is the distance, the faster the reconfiguration and the processing time, and the lower the network latency. Furthermore, if the functionality of high-speed switching is scalable with signal capacity, the network design can be simplified, and the amount and the quality of service delivery can be improved.

6.2.3 OEO ROADM Node

OEO ROADM network node is mainly based on an electrical switch (e.g., cross-point switch) connected with WDM optical transceivers (Tx/Rx in Figure 6.2).

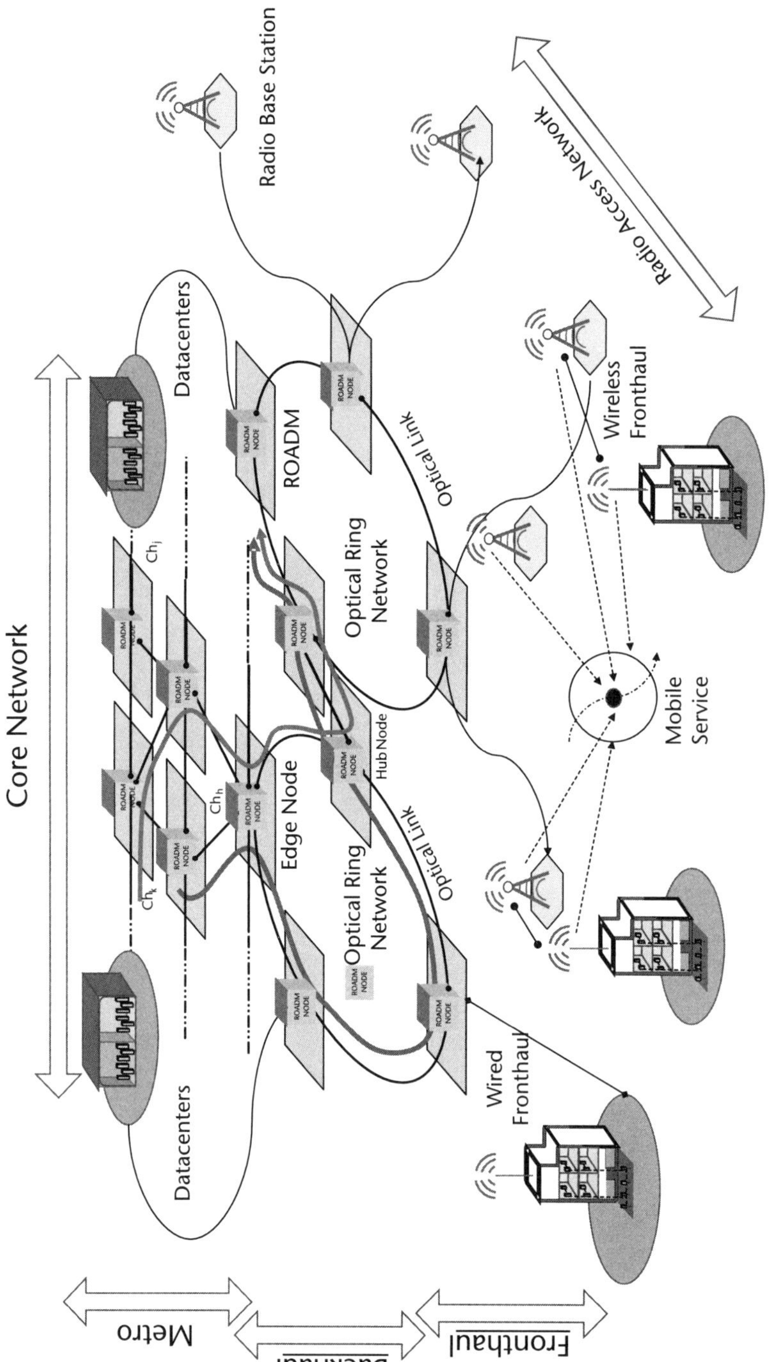

Figure 6.1 Generic RAN structure with optical nodes interconnected delivering services for the RAN.

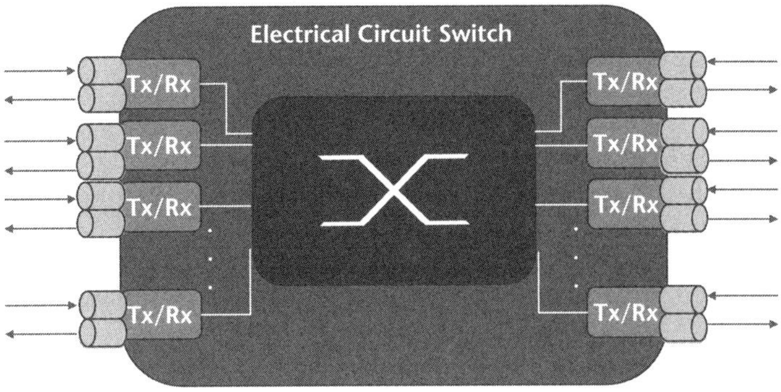

Figure 6.2 General scheme of an OEO node applying the electrical circuit switch to all data flows.

Transceivers are wavelength-specific lasers that convert electrical data signals from data switches into optical signals (e.g., encompassing directly modulated lasers, external modulators, photodiodes); performing the optoelectrical and electro-optical conversion in such a way that the electrical switch can perform WDM routing and activate multiple connections between optical network devices.

The OEO node is capable of perceiving wavelength switching in the electrical domain, with the possibility of regenerating the routed information locally. An OEO ROADM node is a flexible way of obtaining a high-speed, cheap optical network reconfiguration, only if the OEO converter is cost effective. It segregates bypass transit traffic into the optical layer, offloading the router part and consequently reducing its size, power consumption, and cost.

The parallelism in transmission and reception is obtained with an array of N optical electrically driven transmitters and receivers for OEO conversions. Optimization of the network resources can be thus achieved, allowing each node to dynamically use the bandwidth that is available at any time, minimizing the waste of bandwidth and the latency in the network by performing subwavelength routing operations directly in the electrical domain, and preparing transmission in any available network direction toward any available network resources on any available wavelength (i.e., directionless, contentionless and colorless operational mode, respectively). Furthermore, 3R regeneration (Reshaping, Retiming, Reamplifying) at each node makes transmission robust and removes the need for using the FEC technique with reduced overhead and latencies. All the above characteristics scale with channel baud rate in agreement with electronics evolution for the realization of cross-point switches, realizing in the OEO node the circuit switching operation for the network reconfiguration. Currently, electronics scales up to 60 GHz operation, and leveraging on the latest advances following the More than Moore law, a further scaling toward higher baud rates can be expected, but with unclear cost and risk figure of merits, as the evolution depends on a more photonic-integrated design realizing intrachip interconnections in the electronic cross-point switch [4].

As a matter of cost-effectiveness, the optical transmitter can consist of a photonic integrated circuit (PIC) with N fixed lasers at different wavelengths in case of WDM systems. A wavelength multiplexer combines all the signals on a single output fiber. The optical receiver can be also a photonic integrated device, including a wavelength demultiplexer to separate wavelengths received at the input fiber.

An array of N photodiodes converts the signals from optical to electrical followed by a transimpedance amplifier.

6.2.4 OOO ROADM Node

The OOO node has the ability to perform wavelength switching directly in the optical domain, reducing the number of optoelectrical conversions in the network and aggregating or disaggregating traffic carried on multiple wavelengths. Optical switching can be obtained by using optical fiber switches and/or wavelength selective switches (WSS) [5]. Optical fiber switches are most suitable for those optical access networks where a single information channel is transmitted over a single optical fiber. The OOO ROADM network node, as exemplified in Figure 6.3, is able to select and route the WDM signal wavelength carrying the information between two or more network devices. All-optical wavelength conversion can also be necessary to solve possible contentions occurring when, to maintain full network connectivity, there is no chance to avoid the assignment of the same optical path direction (i.e., the same switch output) to different traffic flows, but being carried on the same wavelength.

As in the OEO node solution, where contentionless operations are avoided in the electrical domain, in the OOO node a possible approach to save the information of conflicting traffic is to optically transfer the signal on another available wavelength, throughout an inline wavelength conversion. The inline wavelength conversion is possible by processing signal spectrum in nonlinear media, at the expense of increased energy consumption and system margins reduction. For instance, the FWM nonlinear effect generated in nonlinear fibers [6], dispersion-shifted fibers [7] or semiconductor optical amplifiers (SOA) [8] can be a suitable solution. In this case, additional light sources should be introduced, requiring more energy consumption, and the optical SNR can be decreased as well, mostly in case optical amplification is used, like in SOAs.

In WDM optical transport networks, all-optical nodes are currently available and are based on the capability of multidirectional switching ROADMs to reroute traffic on every single wavelength directly into the optical domain [9, 10]. Concerning ROADM nodes, operators are expecting an increasing level of flexibility to allow the adding or dropping of any set of wavelengths from any link direction to any add-drop access port [10–12]. As a matter of fact, the dynamic colorless,

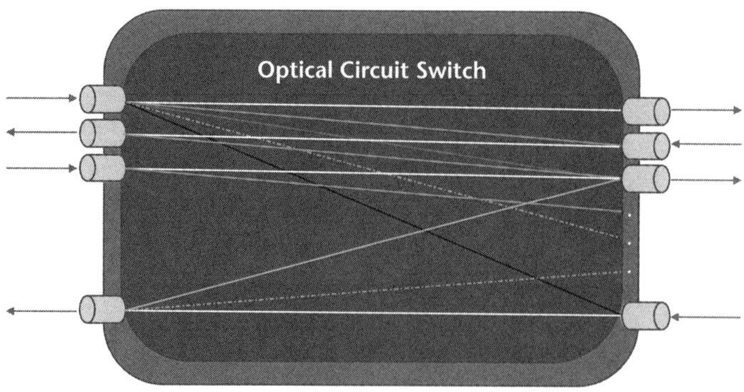

Figure 6.3 General scheme for OOO node applying the all-optical circuit switch to all data flows.

directionless, and contentionless add-drop access will give the operator the possibility to optimize the utilization of resources, eliminate manual intervention, and support rerouting functions in case of faults in a cost-effective way.

Leveraging on silicon photonics integration, it is possible to design a simple ROADM architecture based on a scalable matrix of optical switching elements, allowing for lower power consumption, lower footprint, lower cost, and higher capacity with respect to the state of the art [13, 14]. The envisaged ROADM should be all-optical and designed to efficiently reconfigure the optical path of each WDM channel coming from any direction towards any direction add-drop channels to and from any direction. In integrated and scalable switching matrix architectures, it is desirable to reduce cost, footprint, and power consumption to perform bidirectional switching for the dropping and adding of channels at the same wavelength by the same switching element. Among the possible realizations, those based on the use of microring resonators as switching elements represent a simple and valuable innovative approach. This chapter reports the design tool to properly configure a switching element based on microring resonators by introducing a simple example of an analytical model.

Examples of alternative switching components based on microring resonators, leading to innovative ROADM nodes based on silicon photonics technology, are analyzed in Section 6.3.

Thanks to microrings' characteristics and their use in the design of a switching matrix, a huge number of switching functions can be integrated on a single silicon-based chip, thus realizing a fully operative N×M switching matrix [13]. The cost and the power consumption of a so-fashioned switching matrix can be significantly lower than those realized by cascading several 1×N WSS modules to have the same performance. When designed to have particular configurations (e.g., microrings coupled with parallel or cross-grid waveguides [13, 14]), the microring resonators can simultaneously perform a number of optical functions, like add-drop functions with the bidirectional light-path commutation of signals carried on the same resonating wavelength [14]. The potential to integrate on a single silicon photonic chip all functionalities otherwise performed by interconnecting more than one WSS in a complete switch matrix is evaluated in a different node configuration and network topology in Section 6.4.

6.2.5 OOO SDM ROADM Node

SDM is a technique to increase the overall connection bandwidth by using parallel channels in a single fiber or waveguide being spatially separated, for instance by propagating in different cores or in different orthogonal waveguide modes. Note that the SDM technique is continuously tested by several components and system vendors and operators. A viable solution for site-specific networks still needs a persuasiveness effort by the research community.

Tightly packing spatial-independent channels into a single fiber makes crosstalk between channels an obvious potential problem. Significant crosstalk in a transmission line would have been particularly problematic. However, today, coherent detection systems offer the possibility of electronically removing crosstalk at the receiver, thus giving new attention to the potential of building an innovative network platform to fully exploit MCF, FMF or FM-MCF. The new platform can perform signal multiplexing and switching in the space dimension, adding a further degree of freedom to the OEO and OOO optical nodes discussed above which are operative in the wavelength and time dimensions being capable to commute traffic direction on different wavelengths at different times. Space switching can be realized by optically interconnecting MCF or FMF fiber inputs/outputs with functional modules for channel selection and routing (i.e., free-space, thin-film filters have been demonstrated to demultiplex and switch orthogonal modes [15] and spatial optical backplanes can be demonstrated, too [16]); however, there is a high degree of complexity as a result of the required demultiplexing of the signals transmitted in the various MCF cores or FMF admitted modes, which make the above solutions hardly implementable.

A comparison of wavelength and spatially multiplexed channel traffic routing in an optical network can be found in [5]. Even though spectral allocation strategies result in better network performance, spatial allocation offers the advantage that, by limiting flexibility to a certain extent, can derive cost and benefits from a reduction in the number of required switching elements, without compromising the network operation to an unacceptable degree [5].

6.3 Optical Switching Technologies

The optical switching operations are basically performed by the switching matrix, which is the reconfigurable network element used to operate the traffic routing from one direction port to another direction port, or from one direction port to an add-drop port for local termination or transmission by using directly connected transceivers or transponders. An example of the optical wavelength switching matrix, whose capacity can be expanded by using the SDM technique, is reported in Figure 6.4, defining an optical matrix switch to route traffic from any to any direction, and to locally add-drop traffic.

Two main system blocks can be evidenced: The optical line switching unit (e.g., Ports a, b, c) and the local add-drop unit. The former comprises, for each direction, at least one 1xN (with *N* equal to the node directions, $N = 3$ in Figure 6.4) WSS module to selectively combine onto the output fiber the wavelengths distributed from different matrix switch inputs. Further details on the current state of the art of WSS are reported in Section 6.3.1.

A local add-drop unit includes arrayed waveguide gratings (AWG) [17] to interconnect the optical line switch unit to fixed wavelength transceivers (Tx/Rx modules in Figure 6.4). The array of AWG demultiplexers and multiplexers is used in the scheme of Figure 6.4 to either separate or combine the distributed or aggregated signals to or from each network direction, respectively, before the local receiver or after local generation at the transmitter modules. The AWG module can consist of an array of different length waveguides [17]. Each waveguide is curved and overlays with the neighboring ones, designing superimposed arches in such a way that the inner waveguides are shorter in length than the outer ones.

If an optical WDM comb joins the start of the array, a percentage of the signal travels down each waveguide. Since they are all traveling at the same speed, when they reach the end of the array, they are all staggered slightly in time — the inner waveguide signal arriving first, and the outer one getting through last. These

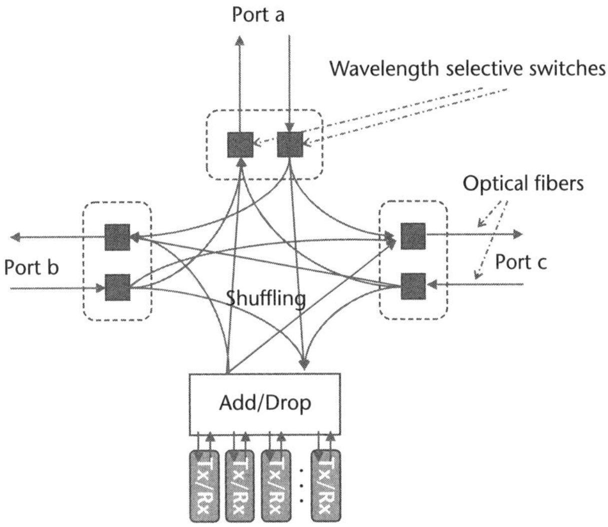

Figure 6.4 Scheme for a ROADM node equipped with WSS and add-drop modules, whose capacity can be expanded by using SDM [5].

time-delayed wavelengths now combine to spit out each individual wavelength in its very own waveguide, cleverly positioned in the only point in space where constructive interference occurs. It should be noted that the number of waveguides used does not have to equal the number of wavelengths that are to be demultiplexed or multiplexed. In general, the more waveguides that are used, the better the AWG can precisely separate each wavelength [17].

6.3.1 Wavelength Selective Switch

The wavelength selective switch (WSS) can be realized by cascading Bragg diffraction gratings fabricated in planar lightwave circuits (PLC) on a unique silicon platform, and exploiting the potential of MEMS (see Figure 6.5) [11]. Those devices operate as waveguide switches and equalizers in a free-space optical matrix (1XN) and can be bidimensional or, more recently, three-dimensional pieces of equipment based on nanotechnology [5]. Unfortunately, they perform optical switching with mechanic inertia constraints, thus practically limiting the switching speed (i.e., by milliseconds) and the optomechanic operations to handle the offloaded packed traffic. An alternative solution also based on nanotechnology is represented by WSS realized by means of LCoS, referring to a silicon chip on which waveguides are coated in liquid crystals, that if properly driven by an electric field causes the change of index of refraction, inducing selective switching on the guided light with a fine tuneable spectrum shaping [12].

The LCoS technology presents significant potentials in switching several optical signals transmitted on different wavelengths with dissimilar bandwidth in a flexible switching matrix 1xN. The coexistence of separated channels with different transmission capacity can be a near future reality by using this kind of devices [18]. Furthermore, the optical signals passing through the network using LCoS may not suffer from the bandwidth limitations introduced by fixed WSS [12]. Due

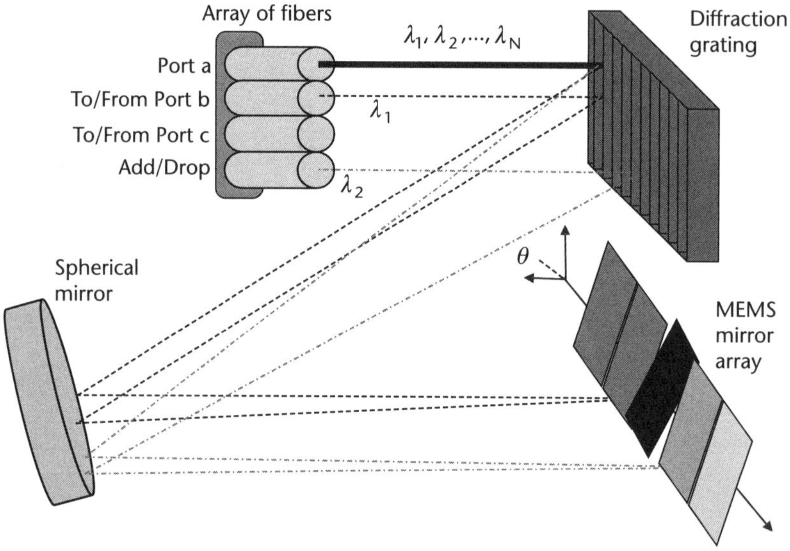

Figure 6.5 Generic scheme and operating principle of a WSS equipped with MEMS [5].

to the complex nature of the control system, to properly drive every single pixel of the LCoS switching matrix, the switching speed cannot be lower than milliseconds.

A standard LCoS WSS system configuration is shown in Figure 6.6, which consists of a linear array of input/output single-mode optical fibers, followed by beam-conditioning optics (fiber-collimating lenses to reduce the beam divergence of the optical fibers, polarization diversity optics for separating each beam into

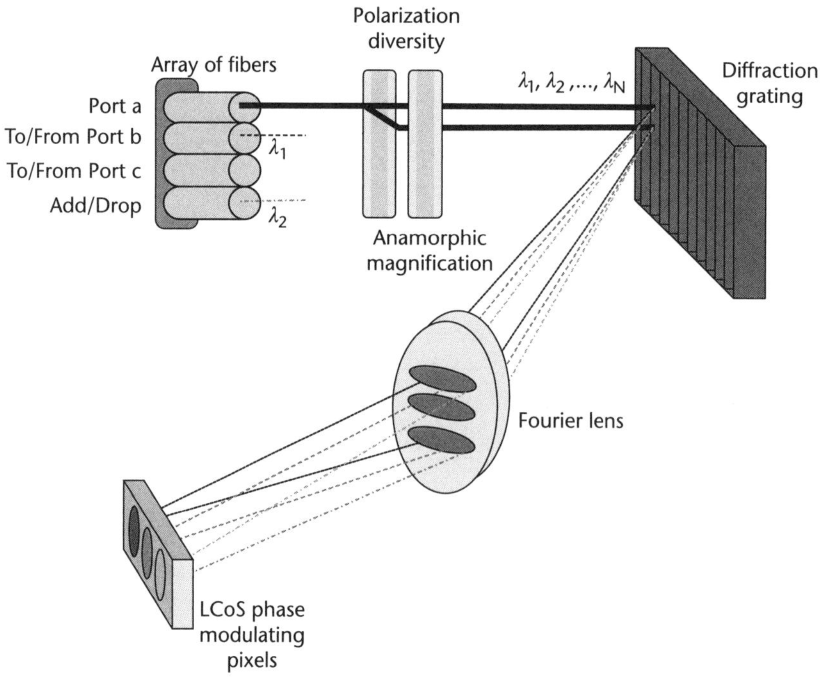

Figure 6.6 Optical arrangement for LCoS-based WSS [5].

two copolarized beams, and anamorphic prisms for beam expansion along one direction to illuminate a wide grating section without increasing module height). Technologies for beam steering include the fine pixels and micromirrors to allow flexible WDM channel allocation

Following the beam-conditioning optics, parallel collimated elliptical beams are projected onto a diffraction grating. An input beam experiences angular dispersion upon diffraction from the grating, which is mapped to spatial dispersion by a Fourier lens. There are many variants to the basic configuration of Figure 6.6 in commercial WSSs, such as using a curved mirror in place of the lens or employing a concave grating. These variants may offer greater compactness or reduce component count, but essentially achieve the same function, namely, to spatially disperse and focus the wavelength channels composing the input beam at the lens' back focal plane. Since all beams emerging from the input and output fibers (had light been introduced from the latter) are parallel up to the Fourier lens, following the Fourier lens they all focus to identical positions on a wavelength basis. Thus, at the Fourier lens' back focal plane, the wavelength channels are dispersed, and the input/output fiber beams are distinguished by their unique incidence angle. At this plane, where the beams are dispersed and ports spatially superimposed, the switching action takes place by beam-steering the input beam toward the desired output fiber on a wavelength basis [5].

6.3.2 NxM Switching Matrix Based on Silicon Photonics

One of the most promising techniques to realize the ROADM node with the architecture in Figure 6.4 can be based on silicon photonics. Silicon photonics has generated exceptional interest as an optoelectronic material because of its promise to make extremely fast and broadband electro-optic devices possible. By optimizing the optical network architecture, new network node realizations and installation can be possible by introducing the latest advance of silicon photonics technology.

The maturity stage demonstrated by the silicon photonics technology can help to design integrated photonic switch matrixes with optimized power consumption and simplified design (i.e., drastically reducing the number of optical elements to be used in the node architecture), while still ensuring large system margins and full network reconfigurability.

Other alternative technologies can be evaluated to optimize energy and cost in new telecommunication network systems, and can be based on different materials than silicon. A comparison of silicon potential with the possible solutions based on III-V and bidimensional materials (e.g., graphene) in the definition of new optical switch matrices is reported in Appendix 6A.

Instead of using several WSS and AWGs to realize a complete switch matrix, silicon photonics allows integration of all needed functionalities within a single optical element. The NxM switch matrix can be designed to shuffle, on request, the connections among the different node directions and add-drop ports (see Figure 6.4).

The matrix dimensions in a ROADM node depends on the number of wavelengths, of node directions, and of connected transceivers, where p is the different directions (e.g., typically 2 or 4), *w* the different wavelengths (e.g., in the C-band, WDM networks have 48 wavelengths spaced 100 GHz apart), *b* the transceivers connected locally (e.g., typically 30% of the number of wavelengths traversing the node), *N* equals *p* multiplied by *w*, and *M* equals b [13, 14].

To enable high-speed signal processing with an optical NxM switching matrix, silicon photonics-based solutions are explored in the following section.

6.3.3 CMOS Photonics for the ROADM Node

The integration of an optical wavelength selective switching matrix over a silicon wafer through CMOS technology is the considered technology for the reduction of needed elements to perform network node reconfiguration [13, 14].

CMOS technology has allowed an unprecedented reduction in footprint of waveguides, and microring resonators are a prime realization example, representing microring as a simple scheme of a basic wavelength selective switch element. The microring resonator consists of a ring-shaped waveguide such that resonance occurs when the optical path length of the resonator is exactly a whole number of wavelengths. Microring resonators, therefore, support multiple resonances, and the spacing between these resonances, the free spectral range (FSR), depends on the resonator length. The microring resonator can be coupled with one or two adjacent straight waveguides (e.g., positioned in parallel or in cross-grid configuration [14]), when the traversing light spectrum matches with one or more multiple resonances. The resonance wavelength depends on the radius and the refractive index of the microring, while the coupling efficiency strictly depends on the proximity of the microring to the waveguide, hence on its design [14].

The microring resonance can be controlled thermally. The temperature variation causes the refractive index to change in the microring waveguide and so does the resonance wavelength. The microring resonator can be therefore enabled to transfer the resonant wavelength energy from a straight-through waveguide to a sufficiently-close adjacent one, to add or drop a signal locally in the optical node. The switch element can be disabled by setting the microring temperature out of resonance. The resonance thermal change dynamics are in the order of microseconds [13, 14].

Another alternative microring driving technique can be set to perform faster switch (i.e., switching time in nanoseconds) and can be done by means of a CMOS process with a silicon–SiO2–metal gate structure acting as a capacitor on a ring waveguide [19].

A layer of metal covers a silicon ring-shaped resonator, modifying its properties. The capacitive actuation of the switching functionality by a voltage-driven change of the refractive index in the microring resonating waveguide is the basic operation. The refractive index change is due to local density carrier variation in the silicon waveguide. The proposed capacitive solution overcomes the need of continuous heating of the device to keep the microring filters tuned in/out of the resonance and therefore reduces 'in operation' energy consumption, but introduces significant optical loss increment which may strongly influence the microring resonator performance in terms of wavelength selectivity range and bandwidth.

6.3.4 Switching Element Design

The switching element characteristics must be carefully designed to reduce the adjacent channel crosstalk issues when operating in a high order cross connect switch for ultradense interconnectivity, in case the switch element performs tight filtering on the selected wavelength. For instance, effective technological implementation can be based on a matrix of microring resonators performing selective spatial switching at different wavelengths with microseconds time response. The microrings are realized on silicon and can be driven thermally or by carrier injection performing potential wavelength selection and switching [13, 14, 19].

The microring resonator and its possible use in an optical switch matrix are extensively described in [13, 14] and consist of a resonant cavity made of a ring waveguide coupled with one or two straight waveguides positioned in proximity.

Each microring must be designed with a diameter that meets the following equation with the resonant wavelength λ_{res}: [19]

$$2\pi n_{eff} r = m \cdot \lambda_{res} \tag{6.1}$$

where r is the microring radius, n_{eff} is the effective index of refraction, m determines the multiple resonances (see Figure 6.7) and is an integer 1, 2, 3....

Moreover, the microring resonators, like other resonant structures, have a periodic transfer function and can be characterized by the full-width half maximum (FWHM) and the FSR. The FWHM defines the bandwidth of the resonance as a function of the coefficient t, the transmission coefficient on the coupled waveguide [see Figure 6.8(a)], and the single-pass amplitude transmission a, including both propagation in the ring and coupling losses ($a = \exp(-\alpha 2\pi r)$ where α is the loss coefficient expressed in [1/cm]) [19],

$$FWHM = (1 - ta)(\lambda_{res})2 \big/ \left(2\pi^2 n_g r \sqrt{(ta)}\right) \tag{6.2}$$

where t is the transmission coefficient in case of all-pass microring resonator, and the product of the transmission coefficients t_1 and t_2 in the case of add-drop microring resonator (see Figure 6.8).

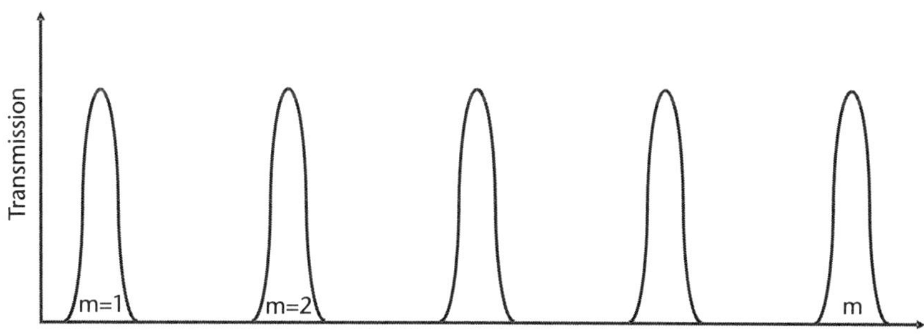

Figure 6.7 The FSR versus wavelength. Resonances of a microring resonator are labeled by the value of m.

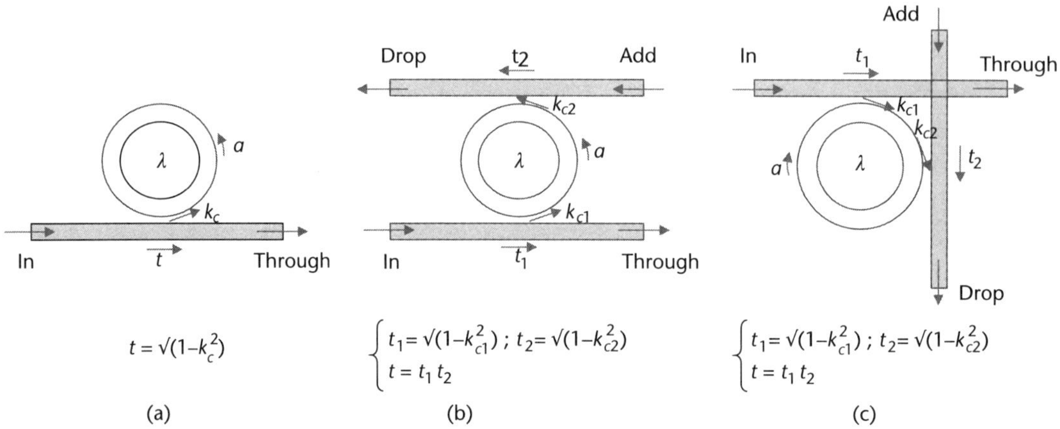

Figure 6.8 Microring resonator design: The coefficient t and the loss a, and the coupling coefficient kc in the case of (a) all-pass, (b) add-drop ring resonator with parallel waveguides, and (c) cross-grid configuration [14, 19].

FSR is the wavelength range between two resonances in function of the wavelength, λ [19],

$$FSR = \lambda^2 / (2\pi n_g r) \tag{6.3}$$

where n_g is defined as the group index and is not an effective index, but takes into account the dispersion of the waveguide. The group index and the corresponding group velocity $v_g = c/n$ describes the velocity at which the envelope of a propagating pulse travels, and is a characteristic of a dispersive waveguide [19]:

$$n_g = n_{eff} - \lambda_0 \left(dn_{eff} / d\lambda_0 \right) \tag{6.4}$$

where λ_0 is the wavelength in vacuum.

The ring resonator FSR is due to the resonating structure of the ring waveguide. By defining the radius of the ring and the coupling coefficient with the straight waveguide in proximity, the resonating wavelength is defined together with the optical bandwidth, and so the FSR by increasing the microring radius — the smaller the FSR, the higher is the microring FWHM—to the expense of higher coupling and propagation losses.

The quality factor of a microring resonator can be represented by the Q factor defining the spectral range of the resonance condition for any value of the microring radius

$$Q = \lambda_{res} / FWHM \tag{6.5}$$

High Q factor can be usually achieved by strongly reducing losses. This target strongly depends on the fabrication process uniformity and circuit/device density over a silicon wafer (e.g., 200- or 300-mm wafer scale production).

An example of microring implementation in a switch matrix with an optimized Q factor is reported in Appendix 6B.

6.4 Application Examples of the Silicon Photonics Integrated ROADM Node

A practical approach to fabricate a microring resonator on silicon with CMOS process on a silicon wafer, aiming to avoid a tight fabrication tolerance of the ring diameter, is to set FSR at a value no larger than about tens nanometers; this results in a ring diameter of about tens micrometers in line with expected tolerance of an existing CMOS production process [13]. Possible matrix schemes achievable with a microring resonators are reported in [14]. The idea is to integrate on a single wafer all those switching elements and broadband transceivers to complete the optical node design for future feature-rich 5G networks; lower power consumption, lower footprint, lower cost and higher capacity with respect to the current state of the art. The optical node may have N optical ports dedicated to offload traffic from one direction to any other, and M add-drop ports exploited to aggregate or disaggregate traffic over any available wavelength [13].

The WSS elements are the microring resonators. Each single matrix row has microring resonators with the same resonating wavelength. Each microring resonator can be in two possible states to either enable or disable optical path commutation — the ON and OFF state respectively. Local transceivers can add and drop traffic on any available wavelength in any direction [13].

With reference to the basic node architecture of Figure 6.4, Figure 6.9 reports a possible switch matrix realization, encompassing microring resonators and AWGs, which can be fabricated at wafer level in CMOS technology on board of a silicon photonic chip. Expansion of switch matrix details is in the insets on the right part

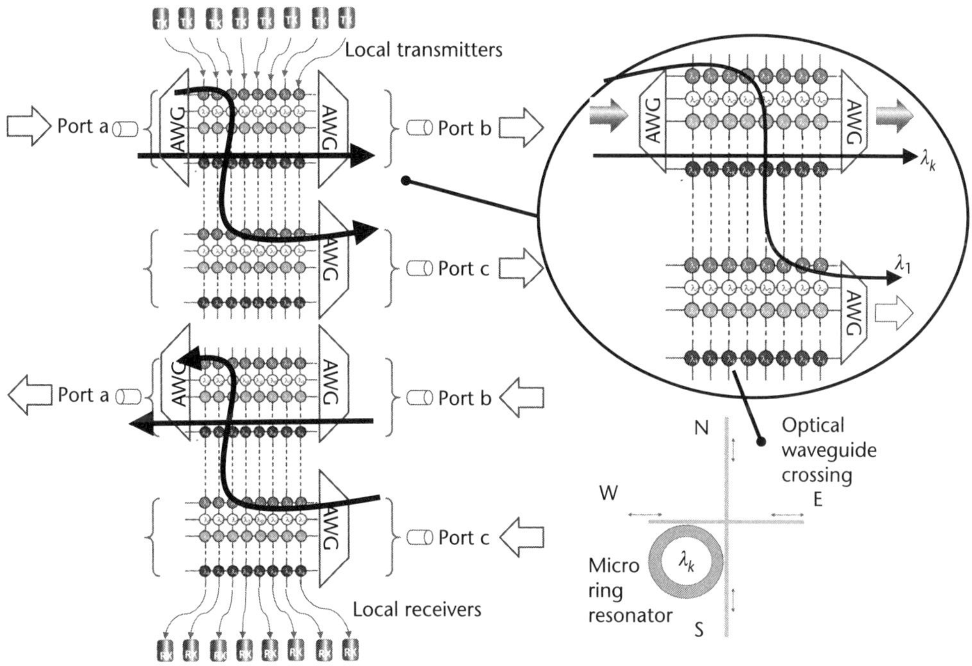

Figure 6.9 ROADM node with N × M switching matrix. The microring resonators onboard are designed in the cross-grid configuration so to transfer the resonating wavelength from one row on the adjacent column.

of Figure 6.9. In Figure 6.9, the microring resonators are designed in the cross-grid configuration (see Appendix 6B) so to transfer the resonating wavelength from one matrix row on the adjacent column when a wavelength routing selection is performed. The reported example is about the execution of a possible change of direction (i.e., "from input port a to output port c" instead of "from input port a to output port b") in the optical path assigned to λ_1 while maintaining unchanged the optical path for a previously copropagating wavelength λ_k (i.e., still from "input port a to output port b"). Being the communication bidirectional for all wavelengths, using two parallel fibers at any node port, the counter direction settings for the switch matrix are updated accordingly: λ_1 goes "from input port c to output port a" and not "from input port b to output port a," which still occurs for λ_k.

As a further reconfiguration example of the switch matrix, add-drop operations can be enabled on λ_1. The microring resonators previously enabled to perform the path direction change for λ_1 are now put out of resonance. If λ_1 is locally dropped and added by the *m*th local receiver and transmitter, the microrings resonating on λ_1 are now both positioned on the *m*th column, but on the first rows corresponding to "input port a" and "output port c," respectively. "Input port c" and "output port a" are no longer affected by the add-drop operation being locally performed by the optical node. All microring controls can be performed by tuning the working temperature of every single switch element involved.

6.4.1 Simplified Silicon Photonic ROADM

The ROADM outlined in Figure 6.9 and the other possible solutions reported in [5] are designed to operate in metro and core networks to have full reconfigurable architecture. Among the most critical parameters to be taken into account is the optical loss due to the insertion of all those optical elements (e.g., AWG, WSS, optical filters) to complete the ROADM node. To compensate for the optical loss, optical amplifiers are used, at least two for every direction of the optical node. The need for optical amplification, therefore, increases with the number of node directions and requires additional energy supply and costs.

In the aim to optimize power consumption and the overall system cost, optical network nodes can be interconnected by implementing simplified network topologies requiring a limited number of directions. An improved node design leveraging on new photonic technologies, mainly to reduce the optical loss, can be a further valuable contribution.

To address the next evolution of RAN, reconfigurable optical interconnections should be defined in both backhaul and fronthaul networks, but seeking for more power efficient and simplified architectures. In the case of next-generation RAN, the most network-stringent requirements are the reconfiguration time speed and the increment of the channel bandwidth by a factor of 10 (from 10–40 to 100–400 Gb/s), for the delivery of new time critical services (i.e., below 1 ms) and for the availability of the network resources to the final user seamlessly.

The typical optical access node can be designed with a lower number of optical input/output ports (i.e., node directions) with respect to metro and core network examples, which are equipped with complex and expensive optical transport gear to bridge long distances.

In the access network, the functional network topologies are used to reduce the number of optical paths (e.g., interconnecting fibers) and to properly serve the network in a localized area (e.g., metropolitan or regional terminal nodes). Among the possible solutions, the optical nodes can be arranged in a ring network topology. Every node is bidirectionally interconnected with two adjacent optical nodes using a pair of optical fiber, one for each direction.

6.4.2 Simplified Silicon Photonic ROADM Node in Optical Ring Topology Networks

The switch matrix reported in Figure 6.9 requires high optical circuit density and involves the high precision capability of the fabrication process and extensive wavelength uniformity over the silicon wafer to guarantee acceptable Q factor for all ring resonators. Nonuniformity can be induced by the modification of the width and the height of the waveguide and the silicon-on-insulator (SOI) thickness when realizing the microring resonators. This nonuniformity is a crucial problem to be compensated for when the process is completed, by readjusting the resonance of each ring resonator with thermal tuning or carrier injection solutions [20].

A ROADM node based on silicon photonics with a reduced circuit density can be realized with more reliable Q factors, as better wavelength uniformity can be achieved and can become the preferred option for cost-effective installation in the RAN.

If the wanted level of production process precision is achieved, then the microring resonances can be efficiently set and add-drop operation can be more efficiently performed in all possible node direction.

In the ring topology network, the optical nodes route traffic locally by dropping and adding wavelengths from and to a single fiber interconnection, and the communication in the ring can be clockwise or counterclockwise. The plugged transceivers transmission can take place only in one direction and modulated wavelengths are bound to a specific node input/output port [14], with no chance to be routed in the other direction. These boundary connections have no significant impact on the network reconfigurability, since in the ring topology allows for only one of the two possible communication directions among adjacent nodes. A possible scheme for a ROADM node in a ring network can be the one depicted in Figure 6.10, where a simplified structure of the switch matrix enables a better wavelength uniformity in the production process, sustainable power consumption, and cost.

6.4.3 Simplified Silicon Photonic ROADM Node for the Edge Node Interconnection

A flexible architecture can be still designed without the need to introduce complex design, in case more than one communication direction is needed in the access network. This is the case of an edge node, where different networks can converge requiring at least two communication directions. More flexible node design (e.g., quasi-wavelength-direction-bound) can be implemented by exploiting some functional realization of the ring resonators to have them fast tuneable on different optical wavelengths with good precision.

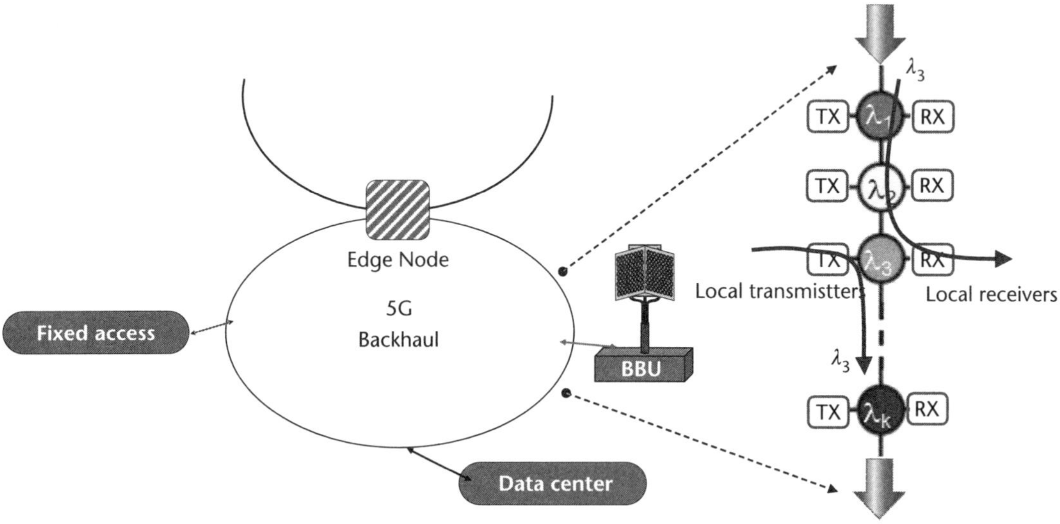

Figure 6.10 ROADM node equipped with a simplified switch matrix for add-drop operation performed in the same WSS element.

The limits of production precision in wavelength resonance, $\Delta\lambda_{res}$, can differ with the radius and the index errors in the fabricated waveguide.

$$\Delta\lambda_{res} = \Delta n_{eff} \cdot 2\pi r/m \tag{6.6}$$

Typically, $2\pi r/m$ is equal to 0.628 for a ring with a radius $R = 10$ µm and a resonant wavelength of $m = 100$ around 1.55 μm (i.e., central wavelength in the C band of the telecom spectrum). A nonuniformity of a few nanometres of the height of the ring waveguide may theoretically cause 1% of Δn_{eff} change and consequently, the resonant wavelength can be shifted by nearly 6.3 nm with respect to expectations [21]. The need for calibrating every single microring in the production phase to compensate for the unwanted resonance shift is crucial and requires time to finalize the entire process. The higher the number of cascaded microring resonators, the longer the calibration process, and the higher is the production cost.

A possible technique to compensate for the nonuniformity of resonating wavelength can be the increment of the bandwidth of a ring, so that a common wavelength can be more easily found among different rings as reported in [20]. The solution increases the interchannel crosstalk [14], but the issue can be solved by increasing the wavelength spacing between adjacent channels. This is noteworthy since as the Q factor decreases, the FSR also decreases, and the optical spectrum increment due to the augmented channel spacing should not be wider than the FSR, otherwise multiple channels can resonate through the same microring. An example of wavelength distribution for a given FSR is reported in Figure 6.11.

The need to make use of a limited FSR to bring the production costs down could bring around the adoption of a novel ROADM that is beneficial for using the same switching elements for different optical nodes in all access networks topologies. This can be done with a carefully designed periodic transmission transfer function of the microring resonators. With silicon photonics microrings, it is pos-

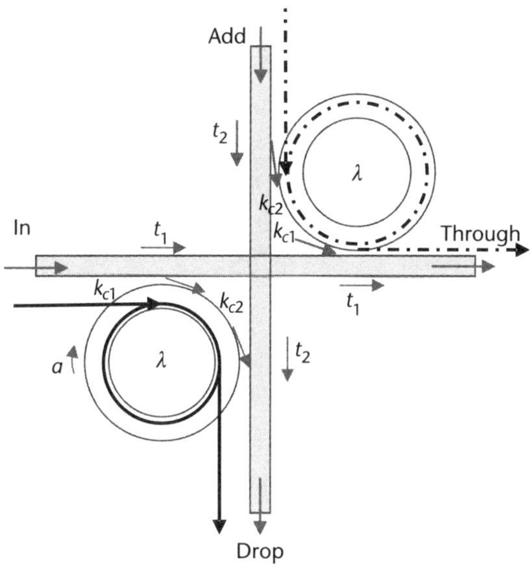

Figure 6.11 Periodic resonance in quasi-color-bound ROADM. FSR is wider than the total wavelength range used in the access network where the ROADM is installed (see Appendix 6B).

sible to achieve tuning of a few nanometers by slightly changing the temperature or the number of injected carriers [21].

Given the above basic characteristics of microring production strategy with increased wavelength uniformity and tunability, a new functional version of the ROADM node can be proposed and can be suitable to serve in the edge node of an access network and can have the architecture reported in Figure 6.12.

In this system implementation, with the ROADM capable of working in multiple regions of the spectrum, the add-drop ports are quasi-color bound (with the possibility to switch wavelength at multiple FSR, too).

The proposed architecture requires a limited number of wavelength switches, as each microring resonator with an extended wavelength uniformity can be tuned within several nanometers in the optical spectrum to select different wavelengths in a WDM comb, thus increasing the possibilities of switch matrix reconfiguration, and reducing the number of rows and columns needed to achieve full connectivity. Practically, each connected transceiver can add and drop traffic in any node direction.

6.4.4 Simplified Silicon Photonic ROADM Node for Fronthaul Networks

In a typical fronthaul configuration hub-and-spoke network architecture [22], the switch matrix can be further reduced, as only two possible counter directions are envisaged for communication among the nodes. The broadband transceivers (e.g., 100–400 Gbit/s) are the optical interfaces connecting an edge router in the hub node to the baseband units (BBU) of the radio base stations to process the radio signal to be transmitted toward the remote radio units (RRU) [22]. The ring resonators can be tuned on all wavelengths assigned to the fronthaul network for interconnection of RUs. A so-fashioned optical node can be designed as in Figure 6.13, where interconnections can be reconfigured by properly driving two rows of

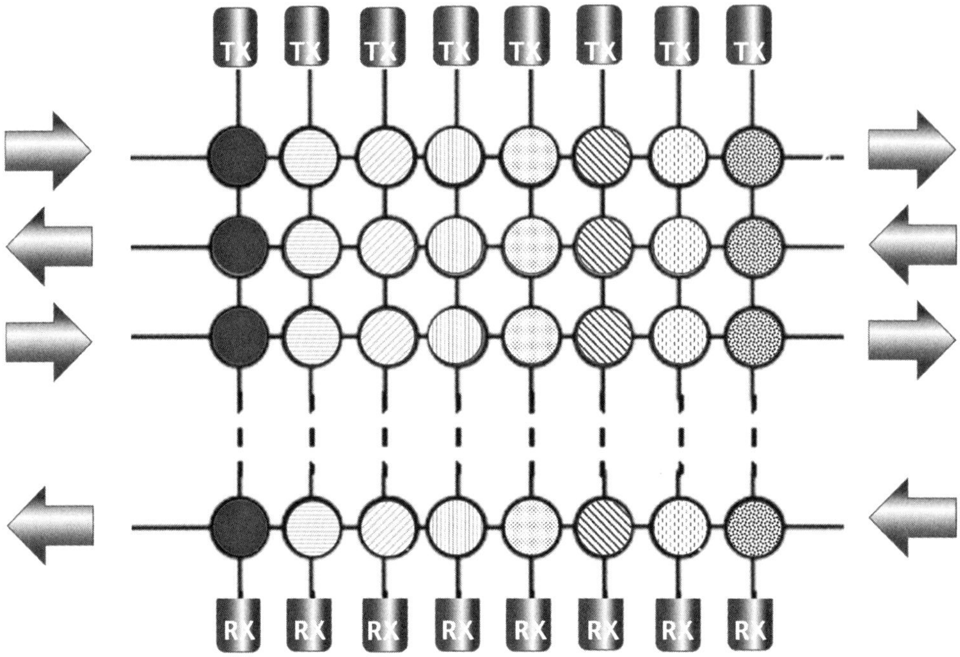

Figure 6.12 Simplified version of an optical ROADM node. Each microring is tunable to select different wavelengths, and can drop and add wavelengths from any to any directions, respectively.

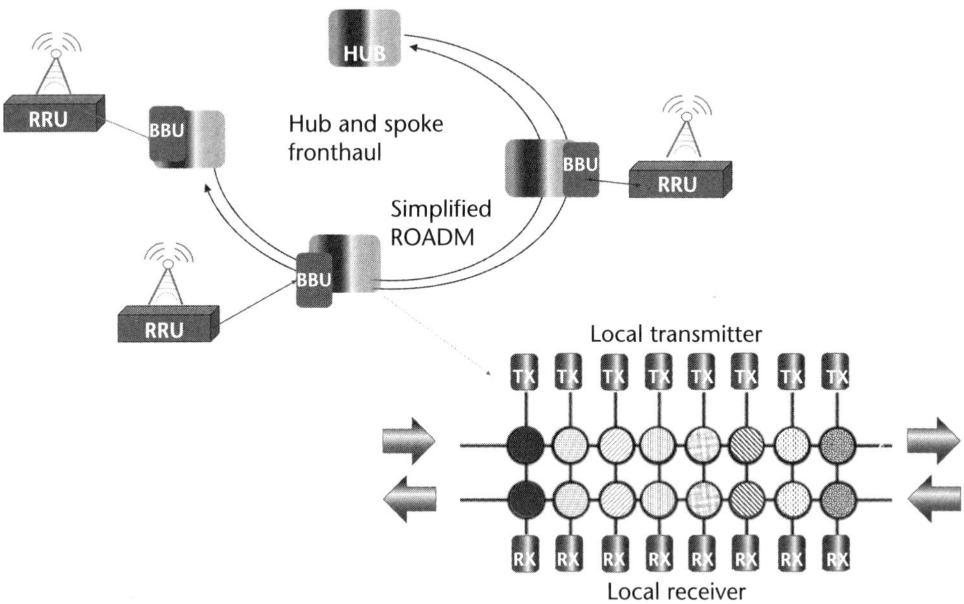

Figure 6.13 Simplified ROADM for hub-and-spoke network topology for fronthaul networks.

tuneable microring resonators for bidirectional transmission, as reported in [22], where less than 1 pJ/bit energy consumption is demonstrated.

6.5 Conclusions

Starting from the state of the art of optical switches leveraging on micromechanics or liquid crystal over silicon pixels, further research and development works are leading to functional solutions for next-generation RAN, where for a matter of network resource use optimization and latency reduction offload of traffic in reconfigurable optical communications is needed. The definition of simplified versions of ROADM node operatives in the metro and core network for access network applications has been described, mainly using silicon photonics technology. Further research and development activity is needed to bring to maturity the silicon photonics, also exploiting the basic physical characteristics of other materials like III-V semiconductor materials. Advanced study on the use of super-metal and polymers will lead to better performing optical systems coping with the need for fast network reconfigurability and network use optimization.

References

[1] Ericsson Mobility Report, November 2018; https://www.ericsson.com/en/mobility-report/reports/november-2018.

[2] Gerstel, O., M. Jinno, A. Lord, and S. J. Ben Yoo, "Elastic Optical Networking: A New Dawn for the Optical Layer?" *IEEE Commun. Mag.*, Vol. 50, No. 2, 2012, pp. s12–s20.

[3] Tomkos,I., S. Azodolmolky, J. Sole-Pareta, D. Careglio, and E. Palkopoulou, "A Tutorial on the Flexible Optical Networking Paradigm: State-of-the-Art, Trends, and Research Challenges," *Proc. IEEE,* Vol. 102, No. 9, 2014, pp. 1317–1337.

[4] Arden, W., Brillouet, M., Cogez, P., et al. "More-than-Moore,",white paper, The International Technology Roadmap for Semiconductor, 2010.

[5] Marom, D. M, P. D. Colbourne, A. D'errico, et al., "Survey of Photonic Switching Architectures and Technologies in Support of Spatially and Spectrally Flexible Optical Networking" [Invited], *J. Opt. Commun. Netw.* Vol. 9, No. 1, 2017, pp. 1–26.

[6] Ahmad, H., et al., "Wavelength Conversion Based on Four-Wave Mixing in a Highly Nonlinear Fibre in Ring Configuration," *Laser Phys. Lett.,* Vol. 8, No. 10, 2011, p. 742.

[7] Mohammed, N.A., M.M. Ragab, and M.H. Aly, "Four-Wave Mixing Based Wavelength Conversion Using Different Types of Fibres," *Int. J. Eng. Sci. Technol.* Vol. 4, 2012, pp. 324–330.

[8] Geraghty, D. F., et al., "Wavelength Conversion for WDM Communication Systems Using Four-Wave Mixing in Semiconductor Optical Amplifiers," *IEEE J. Sel. Top. Quant.,* Vol. 3, No. 5, October 1997, pp. 1146–1155.

[9] Jensen, R., A. Lord, and N. Parsons, "Colourless, Directionless, Contentionless ROADM Architecture Using Low-Loss Optical Matrix Switches," *ECOC 2010 Proceeding,* paper Mo.2.D.2, 2010.

[10] Gringeri, S., et al., "Flexible Architectures for Optical Transport Nodes and Networks," *IEEE Commun. Mag.* Vol. 48, 2010, pp. 40–50.

[11] Colbourne, P., and B. Collings, "ROADM Switching Technologies," *Optical Fiber Communication Conference*, Optical Society of America, 2011, paper OTuD1.

[12] Klonidis, D., et al., "Enabling Transparent Technologies for the Development of Highly Granular Flexible Optical Cross-Connects," in Proc. 16th Int. Conf. Transparent Optical Networks (ICTON), 2014, paper We.D1.5.

[13] Testa, F., et al., "Design and Implementation of an Integrated Reconfigurable Silicon Photonic Switch Matrix in IRIS Project," *J. Sel. Topics Quantum Electr.,* Vol. 22, 2016, pp. 155–168.

[14] Pintus, P., et al., "Analysis and Design of Microring-Based Switching Elements in a Silicon Photonic Integrated Transponder Aggregator," *Journal of Lightwave Technology,* Vol. 31, No. 24, December 15, 2013, pp. 3943–3955, doi: 10.1109/JLT.2013.2276852.

[15] Chen, X., et al. "Reception of Mode-Division Multiplexed Superchannel via Few-Mode Compatible Optical Add/Drop Multiplexer," *Opt. Express,* Vol. 20, 2012, pp. 14302–14307.

[16] Amaya, N., et al. "Gridless Optical Networking Field Trial: Flexible Spectrum Switching, Defragmentation and Transport of 10G/40G/100G/555G over 620 km Field Fibre," *Proc. Euro. Conf. Opt. Commun.,* IEEE, 2012, paper Th.3.D.3.

[17] Smit, M. K., and C. van Dam, "Phasar Based WDM-Devices: Principles, Design, and Applications," *IEEE J. Select. Topics Quant.,* Vol. 2 , 1996.

[18] Sambo,N., et al., "Software-Defined Coderate-Adaptive Terabit/S Based on Time-Frequency Packing," in *Proc. Optical Fibre Communication Conf.,* Anaheim, CA, March 2013.

[19] Bogaerts,W., et al. "Silicon Microring Resonators," *Laser Photonics Rev.* Vol. 6, No. 1, 2012, pp. 47–73, DOI 10.1002/lpor.201100017.

[20] Lea, C.- T., and Lin, B.- C., "A New Approach to the Wavelength Nonuniformity Problem of Silicon Photonic Microrings," *IEEE Journal of Lightwave Technology,* Vol 29, No. 17 September 2011.

[21] Chauveau, C., et al., "Study of the Uniformity of 300mm Wafer Through Ring- Resonator Analysis," *2012 International Conference on Photonics in Switching* (PS), Ajaccio, France, 2012, pp. 1–3.

[22] Iovanna, P., et al., "Future Proof Optical Network Infrastructure for 5G Transport," *IEEE/OSA Journal of Optical Communications and Networking,* Vol. 8, No. 12, December 2016, pp. B80–B92.

[23] Subbaraman, H., et al., "Recent Advances in Silicon-Based Passive and Active Optical Interconnects," *Opt. Exp.,* Vol. 23, No. 3, February 2015, pp. 2487–2511.

[24] Romagnoli, M., et al., "Graphene-Based Integrated Photonics for Next-Generation Datacom and Telecom," *Nature Reviews Materials,* Vol. 3, 2018, pp.392–414.

[25] Shacham, A., K. Bergman, and L. Carloni, "On the Design of a Photonic Network-on-Chip," *Proc. 1st Int. Symp. Networks-on-Chip* (NOCS'07), 2007.

[26] Xia, F., et al., "Ultra-Compact High Order Ring Resonator Filters Using Submicron Silicon Photonics Wires for on Chip Optical Interconnects," *Optics Express,* Vol. 15, No. 19, September 17, 2007.

[27] Luo, X., J. Song, S. Feng, et al., "Electro-Optically Tunable Switches with 100 GHz Flat-Top Pass-Band and 45 dB Extinction Ratio Using Silicon High-Order Coupled Microring Resonators for Optical Interconnects," Optical Fiber Communication Conference, OFC 2012, paper OTu2I.2

Appendix 6A: Silicon Comparison with III-V and Bidimensional Material in Photonics

In silicon photonics, silicon and germanium are the two main materials, and have an indirect bandgap preventing monolithic laser integration onto silicon. To the scope, technological alternatives can be found. Among the possible alternatives, III-V materials and nanostructures (e.g., InAs/GaAs or InAs/InP quantum dots) offer a valid alternative. III/V semiconductor nanostructured materials can offer fast, compact and polarization independent photonic circuitry for optical switches. The silicon-based photonic circuitry is polarization dependent. To realize a polarization-independent switch matrix, polarization diversity should be applied by doubling the matrix structure, thus complicating the production process and the control circuitry [14]. Furthermore, high-performance laser light generation in most cases is realized with direct semiconductor III-V materials (e.g., AlGaAs or InGaAs) as well as other optoelectronic components. Therefore, there is a need for the integration of III-V semiconductors on silicon photonic chips, to achieve complex and advanced heterogeneous silicon photonic integrated circuits.

By applying an electric field, the electrorefraction and the index of refraction nonlinearities in III-V nanostructures can be driven to increase the switching efficiency, and eventually to perform wavelength conversion [23] and increase the level of reconfigurability of the ROADM node. For a more functional version of this type of switch, a hybrid integration of InP-based gain blocks can be designed, in order to compensate for both the on-chip loss and the coupling loss with optical fibers [13].

Because III-V materials electron mobility is much higher than that of silicon, III-V materials can be found in high-speed communications devices for switching time below nanoseconds. As in the case of silicon material also with III-V material nanostructures, the index of refraction and the light absorption are related (i.e., Kramers-Kronig relations), a trade-off should be found between the acceptable absorption loss and refractive index variation range.

Finally, further research is focusing on the combination of silicon photonics technology in combination with other technologies, such as polymers, graphene, and nitride material, leading to hybrid interconnect and routing schemes with the enhanced performance [23, 24].

Appendix 6B: Practical Aspects of Photonic Switch Implementation with Microring Resonators

The basic configuration for a photonic wavelength switch element in an optical switch matrix is the one constituted by a single ring in cross grid configuration as shown in Figure 6.7(c), and here with greater detail in Figure 6B.1.

This configuration is affected by two serious drawbacks. The first drawback can be derived when it is necessary to simultaneously add and drop with the same ring resonator (see Figure 6.B1), as it results in detrimental interference, since the

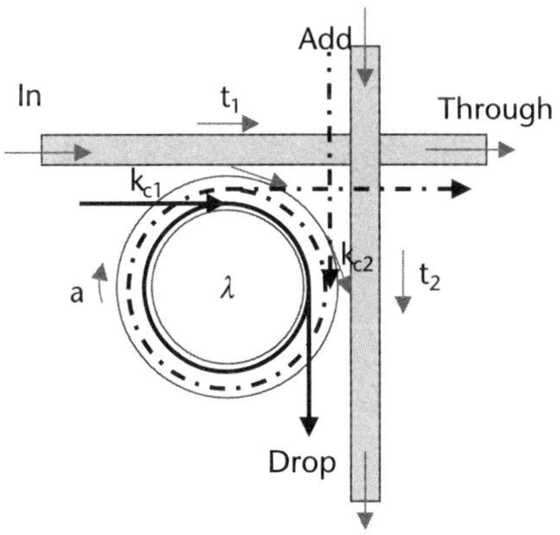

Figure 6B.1 A single ring switch element: Simultaneous add-drop operation.

two signals are copropagating in the ring waveguide and have the same wavelength [14].

The second drawback of single ring configuration is the performance degradation, due to homo-wavelength crosstalk generated by the add-drop crosstalk in the straight crossing waveguides. When a wavelength is dropped, most of its power is coupled to the ring, but a portion of it proceeds on the trough port and disturbs the added signal at the same wavelength, even if with much lower power [14].

To reduce the add-drop crosstalk in a single ring configuration, the only parameter that can be changed is the power coupling coefficient *k*. With a strong coupling, it is possible to reduce the add-drop crosstalk, but at the expense of the filter selectivity. In Figure 6B.2, the transfer function is reported with a strong coupling coefficient, which relates to 30,000 Q factor with a 10 nm ring radius; the homo-wavelength crosstalk can be –30 dB. Unfortunately, in this condition, the filter bandwidth of the drop port becomes too large, as its transfer function is Lorentzian in shape and has too low a out-of-band rejection, and is unable to filter effectively the adjacent channel power. The problem can be mitigated by properly spacing wavelengths in the WDM comb. However, being that the FSR is reduced, the number of available wavelengths is also reduced to avoid simultaneous switching of different wavelengths positioned near the periodic resonance peaks (m=1, 2, ..., n), thus limiting the overall number of channels of the RAN.

Given a value of propagation loss (e.g., 2 dB/cm), the lower that k and Q are, the higher the bandwidth of the drop port and the higher the resiliency to homo-wavelength crosstalk in case of simultaneous add-drop operation in a single microring. The higher that k and Q are, the lower the available number of optical channels in the network. Hence, a trade-off in the design of the switching matrix between crosstalk mitigation and overall network capacity should be found depending on the application.

6B.1 Suitable Microring Configurations

To mitigate crosstalk, alternative designs for the switch element are possible. By doubling the microring resonators in a close grid configuration, as proposed in

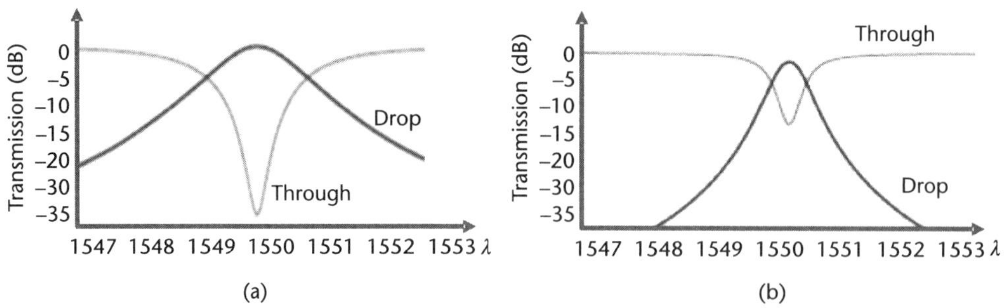

Figure 6B.2 Add-drop and through the port transmission of the microring versus wavelength for two different Q factors: (a) $Q = 3 \cdot 10^4$; (b) $Q = 3 \cdot 10^5$.

[14] in the case of simultaneous add-drop operation, a clear separation of the two optical paths can be achieved, as depicted in Figure 6B.3 [25]. In this case, there is a clear path separation between the signals to be added and dropped on the two different directions, and the detrimental interference is removed.

The double ring cross-grid configuration is beneficial to improve the robustness to the homowavelength crosstalk, keeping high the wavelength selectivity on the drop port to mitigate adjacent channel crosstalk. The residual power of a wavelength to be added, which is not perfectly coupled to the relevant ring in the cross grid, will be no longer mixed with the channel power to be dropped, and vice versa. However, the problem of a Lorentzian shape of the filter transfer function, and the consequent low out-of-band rejection remains even with these two rings in a cross-grid configuration, even if the issue is less stringent with respect to the single microring case.

To obtain at the same time, a large enough flattop bandwidth, a high adjacent channel power rejection, a high add-drop crosstalk rejection, a low loss, and high order coupled microrings structure can be used, as proposed in [26, 27].

As the number of coupled ring increases, the out-of-band rejection increases. The careful and complex design must be made to set the number of coupled rings and the coupling parameters needed to fulfill the specifications of out-of-band rejection, bandwidth flatness, and low add-drop crosstalk in a cross-grid structure. The approach can turn into a yield decrement of the production process, as the match of a high number of parameters should be respected with high precision with regard to the single ring in the cross grid configuration.

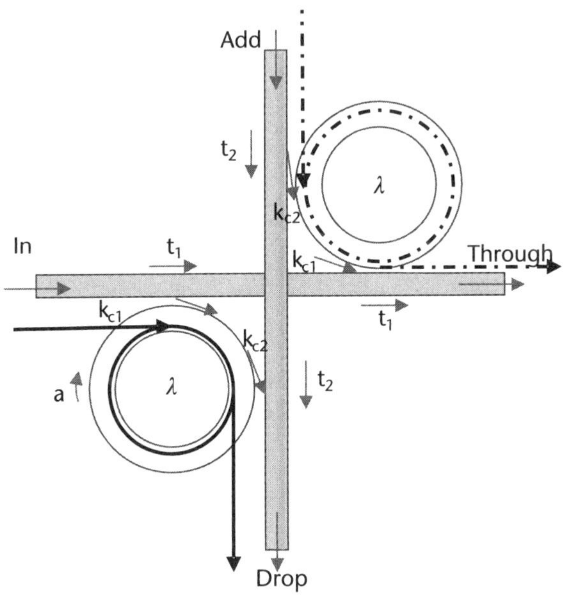

Figure 6B.3 Two rings in the cross-grid configuration: Clear path separation between the wavelengths to be added and dropped [25].

CHAPTER 7

Analog Optical Fronthaul Techniques

7.1 Introduction

Chapter 3 introduced the concept of split RAN architecture, where the different layers of the radio protocols stack are distributed in two or three separate equipment units. In all the split options discussed in Chapter 3, the equipment units are connected by digital communication interfaces. A further analog split option exists, internal to the physical layer, where the RU acts as a mere medium converter, transmitting in fiber a copy of the signal on air, without performing any digital conversion. This split option is referred as analog radio over fiber or, simply, RoF. It simplifies the design of the RU and reduces its power consumption, leading to significant advantages in terms of unit weight, heat dissipation, and reliability. Due to the smaller weight, RoF simplifies the installation and the replacement of RUs placed at the top of antenna towers, leading to a significant reduction of operational costs. However, it is impossible to apply to an analog signal all those processing techniques, like equalization and FEC, that made digital communications so successful. Furthermore, RoF requires a linear channel to avoid waveform distortions; a feature not easy to guarantee in presence of signals with high PAPR, like the ones defined by the LTE and 5G standards. The design tradeoffs of an RoF system, having the basic configuration shown in Figure 7.1, will be the subject of this chapter.

In the RoF system of Figure 7.1, an analog antenna unit (AU) replaces the digital RU. At the AU, the uplink signal is received by the antenna, routed by a circulator or a duplexer to a low-noise amplifier (LNA), and finally, an electrical-to-optical (E/O) conversion is performed onto the signal to then be sent into the fiber link. The E/O conversion is performed by a directly modulated laser (DML) or an external modulator—for example a Mach-Zehnder modulator (MZM) or an electroabsorption modulator (EAM). At either the DU or the CU, at the other end of the fiber link, the optical signal is optical-to-electrical (O/E) converted by a photodiode and sent to a postdetection amplifier (POST) before being sent to the analog-to-digital converter (ADC). In the opposite direction, the downlink signal at the CU or DU is converted from digital to analog, sent into a preamplifier (PRE), E/O converted and transmitted into the fiber link. At the AU, the received optical signal is O/E converted, sent to a Power Amplifier (PA) and routed to the antenna through the duplexer or the circulator. In an RoF system, the distinction between

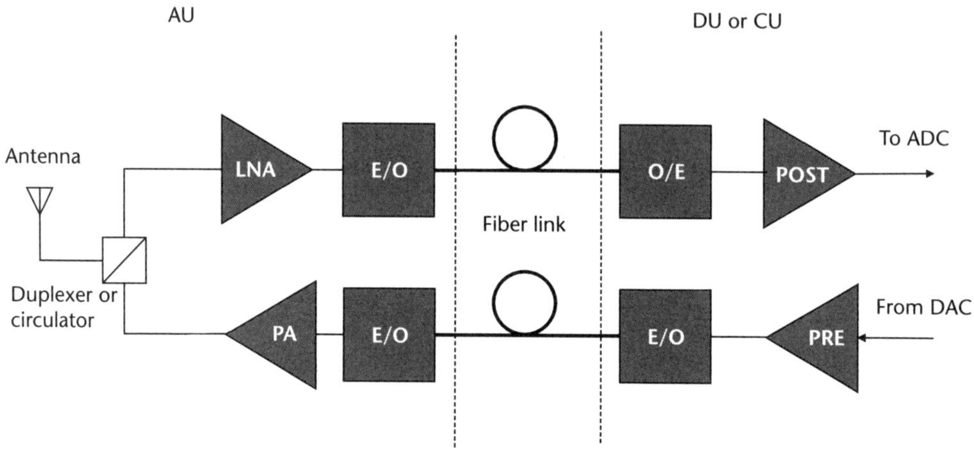

Figure 7.1 RoF system.

CU and DU depends on number and distance of connected AUs. An example of a distributed scenario is a single DU, placed at the base of the antenna tower, connected through short (tens of meters) fiber links to AUs placed at the top of the tower. An example of a centralized scenario is a single CU that serves several AUs, distributed within a radius of a few kilometers.

MIMO adds complexity to the design of an RoF system, due to the high number of antenna elements to be connected. Transmission over parallel fibers (space division multiplexing) is an option only for distributed scenarios over short distance and with a low number of antenna elements. Longer distances and a higher number of antenna elements require frequency multiplexing to reduce the number of fiber cables (of course, TDM is not feasible with analog signals). Frequency multiplexing can be performed in the optical domain (where it is WDM), in the electrical domain (where it is SCM), or in both. WDM has potential cost issues, due to the necessity of using stable lasers emitting at different, nonoverlapping optical frequencies. SCM requires instead intermediate frequency (IF) conversion, performed by means of analog mixers, which increase the power consumption and may introduce nonlinear distortion. Even in the absence of IF conversion, the optical modulation and the photodetection have an intrinsic nonlinearity, which in an RoF system is mitigated using small amplitude modulating signals. This leads to performance limits that will be discussed in the next sections.

7.2 Intensity Modulation of Analog Optical Signals

The electric field of an amplitude-modulated optical signal can be written as:

$$E_{AM}(t) = E_0 \cdot \left[1 + m \cdot v_m(t)\right] \cdot \sin(\omega_0 t + \vartheta) \tag{7.1}$$

where E_0 and ϑ are, respectively, amplitude and phase of the unmodulated electric field, m is the amplitude modulation index, $v_m(t)$ is the modulating signal, and ω_0 is the optical carrier in radians/s. In a photodiode, $E_{AM}(t)$ generates a current I(t) proportional to the instantaneous incident optical power:

$$I(t) = \Re \cdot E_0^2 \cdot \left|1 + 2 \cdot m \cdot v_m(t) + m^2 \cdot v_m^2(t)\right| \tag{7.2}$$

where $\Re$ is the photodiode responsivity. For a small modulation index, $m \cdot max\, |v_m(t)| << 2$, so that

$$I(t) \sim \Re \cdot E_0^2 \cdot \left|1 + 2 \cdot m \cdot v_m(t)\right| \tag{7.3}$$

If the modulation index is chosen so that $m \cdot max\, |v_m(t)| < 0.5$, the argument of the absolute value in (7.3) is always positive and

$$I(t) \sim \Re \cdot E_0^2 + 2 \cdot m \cdot \Re \cdot E_0^2 \cdot v_m(t) \tag{7.4}$$

The constant term in (7.4) is filtered by the electrical frontend of the optical receiver, which is AC-coupled to the photodiode, so that a current proportional to the modulating signal is obtained.

$$I_{AC}(t) \sim 2 \cdot m \cdot P_0 \cdot v_m(t) \tag{7.5}$$

where $P0 = \Re \cdot E_0^2$ is the received optical power. From the previous discussion, it is clear that using a small modulation index is mandatory to neglect the nonlinear distortions due to the quadratic term in (7.2) and to avoid points of singularity caused by changes of sign of the absolute value argument in (7.3). However, a small modulation index has two drawbacks; a significant part of the transmitted power is absorbed by the unmodulated optical carrier, which carries no information, and high signal to noise ratio values are difficult to guarantee; and a high PAPR, which is a critical issue for OFDM and SCM modulating signals that both have high PAPRs. Since both OFDM and SCM are widely used in RoF systems, the PAPR issue will be discussed in more detail in the next section.

7.3 Peak to Average Power Ratio of Multicarrier Signals

In essence, a multicarrier signal can be considered as the sum of N sinusoidal signals, where N is the number of its subcarriers. Putting the amplitude of a single subcarrier equal to 1, the maximum possible amplitude of the sum is N. It can be demonstrated that N is just the amplitude upper bound of an OFDM signal using PSK modulation. However, since independent modulation symbols are transmitted on different subcarriers, the occurrence probability of the maximum PAPR is very low. If M is the number of phase levels of the PSK constellation, only M^2 symbol patterns over a total of M^N will generate the maximum PAPR, with an occurrence probability of M^{2-N}. For example, with N=32 and $M = 4$ (i.e., QPSK), the occurrence probability of the maximum PAPR is 10^{-18}. With QAM, the peak of the multicarrier signal is higher than N and its occurrence probability cannot be calculated exactly. An approximation is obtained considering the baseband OFDM signal of (7.6) as a complex stochastic process [1, 2]:

$$\tilde{s}(t) = x(t) + jy(t) = \sum_{n=1}^{N} a_n \cos \omega_n t + jb_n \sin \omega_n t \tag{7.6}$$

In (7.6), a_n and b_n and the real part and the imaginary part of the QAM constellation symbols, and ωn are the subcarriers frequencies, in radians/s. In presence of a large number of subcarriers, $x(t)$ and $y(t)$ can be modeled as independent, identically distributed, zero-mean Gaussian variables, at any given time t. The average OFDM signal power, P_{av}, is given by the following equalities:

$$P_{av} = x^2 = y^2 = a_n^2 + b_n^2 \tag{7.7}$$

The maximum value of the OFDM signal is instead $max\left[\sqrt{x^2(t) + y^2(t)}\right] = \sqrt{max\left|\widetilde{s(t)}\right|^2}$. Hence, the PAPR can be defined as:

$$PAPR = \frac{max\left|\widetilde{s(t)}\right|^2}{P_{av}} \tag{7.8}$$

Replacing $v_m(t)$ with $max\left|\widetilde{s(t)}\right|$ in (7.2), the photocurrent upper bound can be calculated as:

$$max\left[I_{AC}(t)\right] \leq 2 \cdot m \cdot P_0 \cdot \sqrt{P_{av} \cdot PAPR} \tag{7.9}$$

Equation (7.9) guarantees that the photodiode operates linearly. Similarly, signal upper bounds that guarantee the linear operation of amplifiers, and E/O converters in the RoF system of Figure 7.1, can be expressed as function of the PAPR, and the average signal power at the input of those devices. Since the modulating signal seldom reaches its peak value, it is a frequent design choice to set the modulation index to a value higher than the value strictly allowed by those upper bounds, in such a way that a small outage probability is introduced, but the SNR is improved. To do so, it is necessary to estimate the probability that the signal does not exceed a given threshold, decided by design to keep the outage time acceptably short. Considering $x(t)$ and $y(t)$ in (7.6) as independent Gaussian stochastic processes, $\sqrt{x^2(t) + y^2(t)}$ will have a Rayleigh distribution and the PAPR will be a random variable with cumulative distribution function given by (7.11) [2]:

$$Probability(PAPR < x) = \left(1 - e^{-x}\right)^{\alpha \cdot N} \tag{7.10}$$

The parameter α>1 (e.g., $\alpha = 2.8$) is introduced empirically, without any theoretical justification, to compensate for a mismatch between this simple analytical model and simulations and measurements results. Based on (7.10), the PAPR that is exceeded with probability P is:

$$PAPR = -ln\left[1 - (1 - P)^{\frac{1}{\alpha \cdot N}}\right] \tag{7.11}$$

For example, with $N = 1024$ and $P = 10^{-15}$, PAPR = 4.4, or 6.5 dB. Several techniques have been proposed to mitigate the PAPR in OFDM modulated systems. They will just be mentioned here. Clipping the signal, or limiting its maximum amplitude to a fixed level, is the simplest way to reduce the PAPR. However, it distorts the OFDM waveform, a phenomenon known as self-interference, and increases the out-of-band radiation [1]. Peak windowing [1] reduces the out-of-band radiation, compared to clipping: it consists of multiplying high signal peaks with a properly shaped, nonrectangular window. The window duration should be short enough not to affect many symbols, increasing the BER, but it should also have a narrow bandwidth, not to distort the OFDM spectrum. Cosine and Hamming windows are examples of shapes used for peak windowing. Linear peak cancelation [3, 4] is a technique that uses a time-shifted and scaled reference function, subtracted from the signal, to reduce its peak power. Choosing a reference function with a bandwidth close to that of the transmitted signal, no out-of-band interference is generated. The peak cancellation is performed either in the time domain—processing the signal samples before the DAC—or in the frequency domain, after the IFFT—splitting the signal, processing a copy of it, and subtracting it to the signal on the main path. PAPR reduction codes [5, 6] are coding techniques that act on the modulation symbols and are designed to avoid patterns of symbols that generate signal peaks. Well-designed codes have also error-correction capabilities. Golay complementary sequences [7] are an example of PAPR reduction codes. Finally, with symbol scrambling [8], the input sequence of modulation symbols is scrambled in order to obtain a lower PAPR.

7.4 Optical Modulation in Radio over Fiber Systems

The previous sections discussed the constraints introduced on the modulating signal in an RoF system. This section will focus on optical modulation technologies. In an RoF system, as in any optical transmission system, the modulating signal can be superimposed to the current of a laser diode or applied to the driving voltage of an external modulator, typically a Mach-Zehnder modulator (MZM) or an electroabsorption modulator (EAM).

7.4.1 Directly Modulated Lasers

The operating principle of a laser is sketched in Figure 7.2. Oversimplifying, a laser diode is a cavity obtained by superimposing different layers of semiconductors, designed to obtain three energy levels for photons emitted within the cavity; a fundamental energy level, E_1; an excited energy level, E_3; and a metastable energy level, E_2. Injecting current into the device (a process known as pumping), a photon moves from the fundamental layer to the excited layer, from which it suddenly falls onto the intermediate metastable level. The photon lifetime on the metastable level is long enough to allow the accumulation of many photons before they decay back onto the fundamental state, a condition known as population inversion. However, an incoming photon with a frequency equal to $(E_2–E_1)/h$, where h is Planck's constant, will force the decay of the photons from the metastable level, leading to light emission at such a well-defined frequency. The structure of the semiconductor

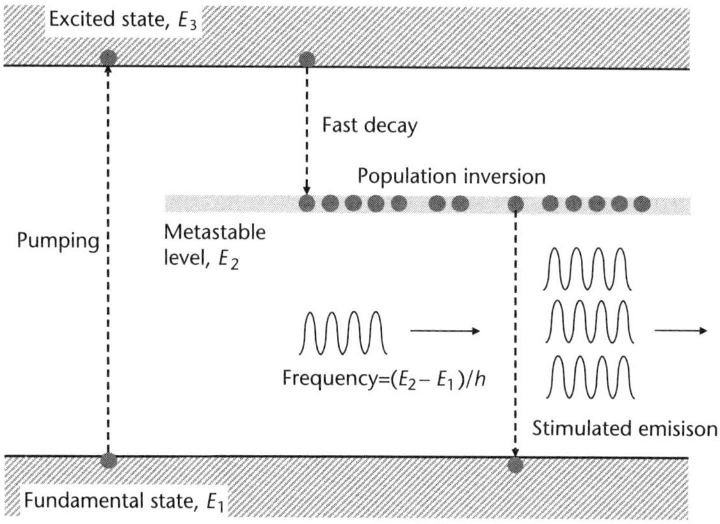

Figure 7.2 Laser working mechanism.

device is designed to confine the photons within the cavity, so that only a small fraction of the photons can escape and the population inversion can be continuously maintained through current pumping.

The evolution over time of the number of photons in the cavity and the number of pump electrons can be calculated by means of the rate equations [9], which are a pair of first order, coupled, differential equations. They depend on device parameters that are difficult to measure, which are usually indirectly estimated from measurements on injected current and optical power [10]. In a directly modulated laser (DML), the modulating signal is applied to the injected current to generate a proportional amount of output optical power. The output power depends on the current as shown in the example of Figure 7.3. The output power is zero until the current reaches a threshold value, sufficient to trigger the population inversion and the stimulated emission. Above the threshold, the optical output power grows linearly with the current, but for high current values, the number of excited photons cannot increase any longer, and the output optical power saturates. The slope of the curve, measured in W/A, is an important design parameter of an RoF system, since it sets the gain of the E/O conversion.

Both threshold current and curve slope are temperature -dependent. The threshold I_{th} grows with the temperature, as:

$$I_{th} = I_{th0} e^{\frac{T - T_0}{T_0}} \tag{7.12}$$

where T is the temperature, T_0 is the characteristic temperature of the device (for example, for a 1,300-nm InGaAsP laser T_0 varies from 50°K –70°K), and I_{th0} is the threshold at T_0. From (7.12), it is obtained $dI/I = dT/T_0$, which means that any current variation is proportional, in percentage, to the absolute variation of temperature. As shown in Figure 7.3, the slope of the curve decreases with the temperature, as well as the E/O gain experienced by the RoF signal. To stabilize the

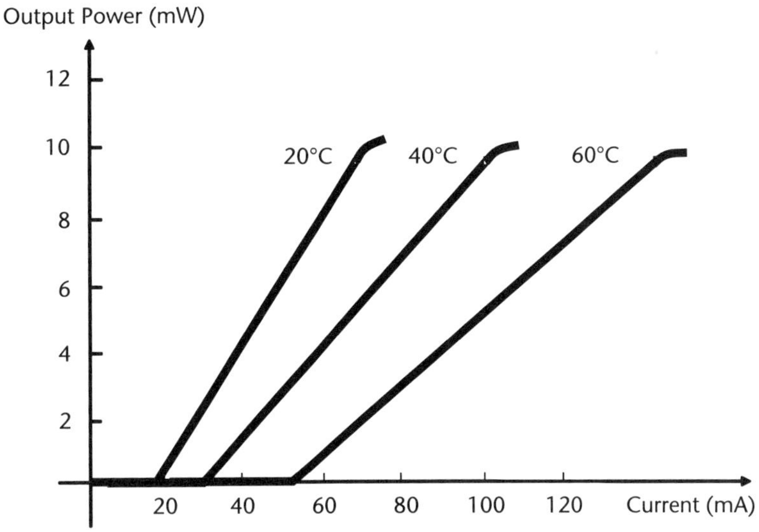

Figure 7.3 Output power versus current in a laser.

gain, thermoelectric coolers (TECs) are used to keep the temperature of the laser constant. In an RoF system, the bias current is set halfway between the threshold current and the saturation current, to allow the maximum amplitude excursion to the modulating signal within the linear segment of the curve. In Figure 7.3, assuming that the laser temperature has been stabilized at 20°C, the threshold current and the saturation currents are 20 and 70 mA, respectively. This gives a bias current of 35 mA, corresponding to 5 mW of mean output optical power.

The curves in Figure 7.3 are valid for static input currents. The dynamic behavior of a DML with small input signals, as in RoF, can be modeled as a second-order low-pass filter [10]. The bandwidth of the filter increases with the bias current, as qualitatively shown in Figure 7.4. In an RoF system, the bandwidth must be higher than the radio frequency carrier.

The small signal response of Figure 7.4 does not apply in principle to input signals with large amplitude excursion, as an OOK digitally-modulated signal did.

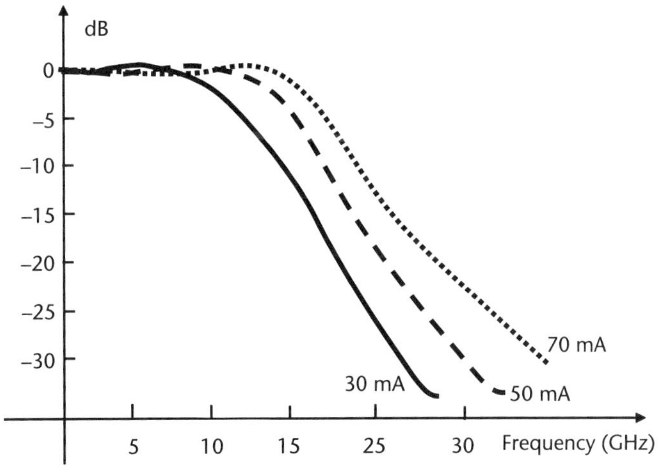

Figure 7.4 DML frequency response with small input signals.

However, it can be demonstrated by manipulating the rate equations that the DML response to a large rectangular pulse still has the typical behavior of a second-order linear filter, with transients characterized by a rise time, an overshoot, and relaxation oscillations (Figure 7.5).

In a DML, any variation of optical power induces variations of the emission frequency. The effect is known as chirp and in an RoF system it leads to undesired signal spectral broadening. In digital transmission systems, chirp may be introduced on purpose to compensate the chromatic dispersion or to generate an FSK signal. FSK optical transmission systems use a chirped DML to generate a quasiconstant amplitude signal with a spectrum having two pronounced frequency peaks [11]. The chirp frequency offset, $\Delta\nu(t)$, from the optical carrier frequency depends on the optical power $P(t)$, according to following equation:

$$\Delta v(t) = \frac{\alpha}{4\pi}\left[\frac{d}{dt} InP(t) + \kappa P(t)\right] \tag{7.13}$$

where α is the chirp factor and κ is the adiabatic chirp coefficient. First and second terms at the right side of (7.13) are called transient chirp and adiabatic chirp, respectively. The transient chirp is significant during the relaxation oscillations, while the adiabatic chirp coefficient is proportional to the gain compression factor [10], which is a coefficient that indicates when the curve in Figure 7.3 starts deviating from the linear shape.

A parameter used in practice to characterize the laser nonlinearity is the 1-dB compression point, defined as the current at which the difference between actual value of output power and extrapolation of the linear segment of the curve in Figure 7.3, is 1 dB. Another parameter of common use is the output third-order intercept point (OIP3). OIP3 is measured feeding the DML (or any other nonlinear system) with two equal sinusoidal tones, at frequencies f_1 and f_2. Increasing the power, P, of the tones, a third tone, called third-order intermodulation product (IM3) will appear at frequency $2 \cdot (f_2 - f_1)$. The power of the intermodulation product increases faster than the signal power. Since it is due to the third-order term of the Taylor expansion around the bias current of the curve in Figure 7.3, it will be proportional

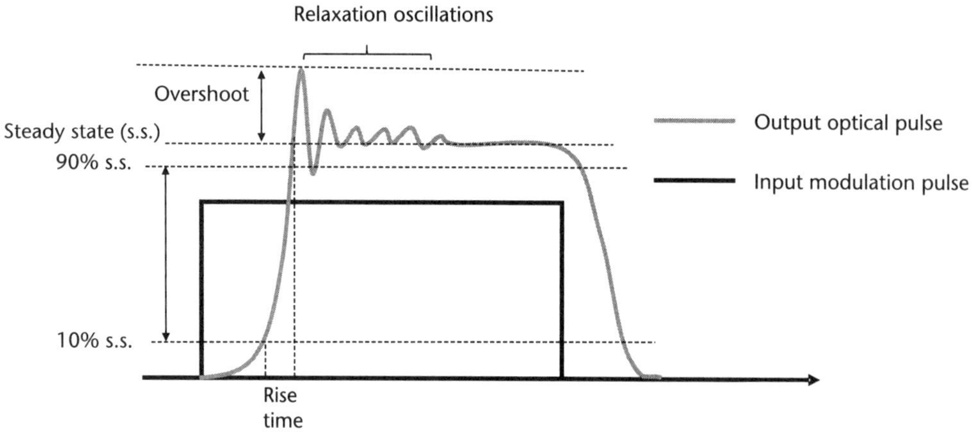

Figure 7.5 DML time response to an input rectangular pulse.

to P^3. Plotting output power versus input power in decibels (Figure 7.6), the input tone will be represented by a straight line with slope equal to one, whereas the intermodulation product will be a straight line with slope equal to three. OIP3 and input third-order intercept points (IIP3) are defined as output power and input power at which signal and IM3 become equal (Figure 7.6).

A similar analysis can be carried out for second-order terms at frequency f_2-f_1. In this case, the slope of the intermodulation product versus the signal will be equal to two rather than three. An analysis of the intermodulation distortion in a DML can be found in [12]. Moreover, composite second-order (CSO) distortion, defined as the ratio of second-order distortion and carrier power [13], is generated by the combined action of chirp and fiber chromatic dispersion.

7.4.2 Mach-Zehnder Modulator

At modulation bandwidths higher than 10 GHz, DML starts suffering from performance impairments. Although DML operation at 40 Gbit/s has been demonstrated up to 10 Km [14], external modulation is the usual choice for bit rates higher than 10 Gbit/s, in digital communication systems, or radio frequency carriers approaching or higher than 10 GHz, in analog RoF systems. MZMs and EAMs are the most used types of external modulator [15]. A MZM is a voltage-controlled interferometer; input light coupled to the device is split into two waveguides, then, the two split signals experience a relative phase shift, obtained by changing the refractive index of the waveguides by means of voltage applied to the two interferometer arms; finally, the two phase shifted signals are coupled again to obtain the output signal.

The working principle of an MZM is shown in Figure 7.7. As consequence of the applied voltages $v_1(t)$ and $v_2(t)$, an unmodulated optical carrier with amplitude E_0 and frequency ω_0 is multiplied by a complex factor y(t) equal to:

$$y(t) = \frac{e^{j\pi\frac{v_1(t)}{V_\pi}} + \gamma_s \cdot e^{j\pi\frac{v_2(t)}{V_\pi}}}{2} \tag{7.14}$$

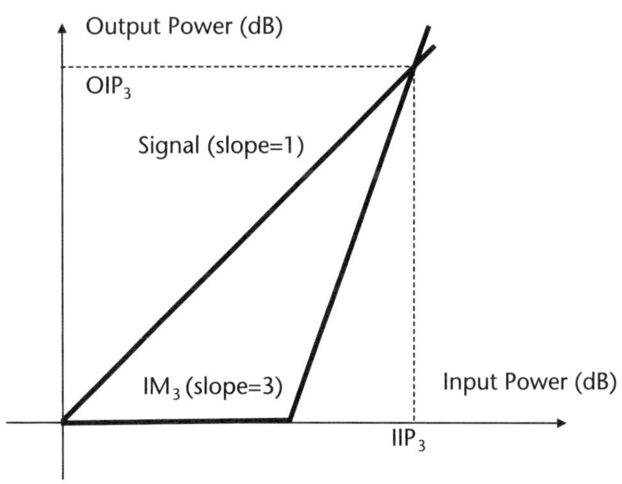

Figure 7.6 Third-order intercept point definition.

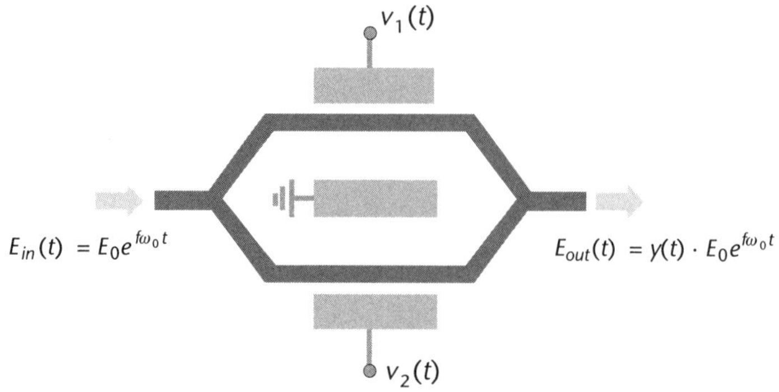

Figure 7.7 Mach-Zehnder modulator.

where $V\pi$ is a constant typical of the device and the factor γ_s accounts for unbalanced or imperfect splitting ratio of the two arms. With a perfectly equal splitting ratio, $\gamma_s = 1$, the following relationships hold between γ_s and extinction ratio, r, defined as ratio of maximum and minimum optical output power:

$$\gamma_s = \frac{\sqrt{r}-1}{\sqrt{r}+1} \tag{7.15}$$

With an MZM, amplitude modulation is obtained by setting

$$v_1(t) = -v_2(t) = \frac{V_\pi}{2}x(t) \tag{7.16}$$

so that:

$$y(t) = \frac{1+\gamma_s}{2}\cos\left[\frac{\pi}{2}x(t)\right] + j\frac{1-\gamma_s}{2}\sin\left[\frac{\pi}{2}x(t)\right] \tag{7.17}$$

If γ_s<1, y(t) is a complex and generates both amplitude and phase modulation. In the ideal case of $\gamma_s = 1$, $y(t)$ is real and no phase modulation occurs:

$$y(t) = \cos\left[\frac{\pi}{2}x(t)\right] \tag{7.18}$$

With reference to Figure 7.7, the output optical power can be written as:

$$P(t) = P_0 \cdot \cos^2\left(\pi\frac{v(t)}{V_\pi}\right) \tag{7.19}$$

where $v(t) = v_1(t)$, and P_0 is the unmodulated optical power, as can be verified putting $v(t) = 0$.

In an RoF system, the voltage is:

$$v(t) = \frac{V_\pi}{2} - m \cdot v_m(t) \tag{7.20}$$

where $V_\pi/2$ is the bias voltage, m is the modulation index, and $v_m(t)$ is the modulating signal, for example, an OFDM signal. For a DML, the slope efficiency, s_{DML}, was defined as the derivative of the power-current curve in Figure 7.3. The same definition applies to an MZM, assuming that the modulating current, $i_m(t)$ is proportional to $v_m(t)$, through the modulator impedance R_m ($v_m = R_m \cdot i_m$):

$$S_{MZM} = \frac{dP}{di_m} = R_m \frac{dP}{dv_m} = \frac{\pi P_0 R_m}{2V_\pi} \tag{7.21}$$

Putting $v_2(t) = \frac{V_\pi}{2} + v_1(t)$ in (7.17), an MZM can be used to generate an optical single sideband (OSSB) modulated signal, whose expression for $\gamma = 1$ is:

$$E_{OSSB}(t) = \frac{E_0}{2} \cdot \left[\cos\left(\omega_0 t + \pi \frac{v_1(t)}{V_\pi}\right) + \cos\left(\omega_0 t + \pi \frac{v_1(t)}{V_\pi} + \frac{\pi}{2}\right)\right] \tag{7.22}$$

To avoid CSO distortion, which is detrimental in an RoF system, it must be $\pi \frac{v_1(t)}{V_\pi} \ll 1$. For example, 20-dB CSO is obtained with $\pi \frac{v_1(t)}{V_\pi} = 0.4$ [15]. OSSB is used to mitigate the CD penalty of RoF systems operating in the C-band [17]. An intuitive explanation is given in Figure 7.8; in an OSSB-modulated signal, independent data is transmitted on each individual subcarrier, whereas in a two-sided amplitude-modulated spectrum, the same data is transmitted on two subcarriers, symmetric around the optical carrier. Chromatic dispersion changes the relative phase of the two subcarriers and, after photodetection, their relative amplitude, leading to a penalty.

7.4.3 Electroabsorption Modulator

The absorption coefficient of a material indicates how deep light at a certain wavelength can penetrate the material before being absorbed. In semiconductors, light with a frequency lower than the band gap does not have enough sufficient energy to excite electrons from the valence band to the conduction band, and is not absorbed, leading to a sharp decrease of the absorption coefficient with the light wavelength. Electroabsorption (EA) is the effect whereby the absorption coefficient changes due to the presence of an applied electric field, as the modulating voltage of an RoF system. EAMs can operate in both C-band and O-band and are realized with waveguides consisting of doped and intrinsic III-V semiconductor layers,

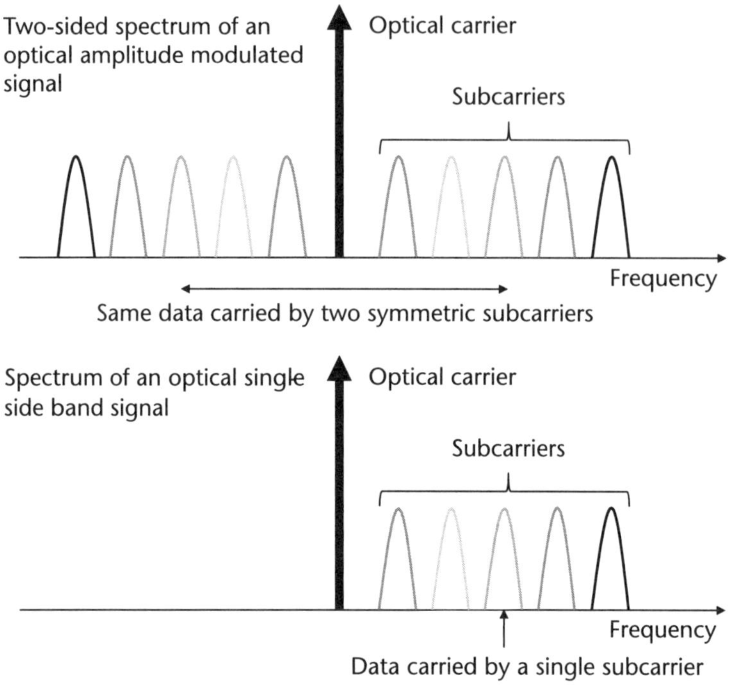

Figure 7.8 Difference between OSSB and amplitude modulated spectrum.

typically alloys like InGaAs, or InGaP, on an InP substrate. The intrinsic layer has a higher refractive index than the surrounding doped layers, to provide vertical optical confinement. Lateral light confinement is achieved by deep mesa etching, in such a way that both light and applied electric field are confined in a small area, to allow for the high intrinsic insertion loss of the device, which is typically in the order of 15–20 dB/mm. The cross section of the main propagation mode in the EAM waveguide is approximately elliptical, with a horizontal axis of 2–4 μm, and a vertical axis of 1–2 μm. The resulting maximum cross section area is 25 μm^2, much smaller than the cross section of the fundamental propagation mode in fiber, which is quasicircular, with an area of about 80 μm^2. Microlens or tapers are used to reduce the resulting coupling loss at the interface between EAM and optical fiber. An EAM exploits two types of electroabsorption effects: the Franz-Keldysh effect and the quantum-confined Stark effect. The Franz-Keldysh effect working principle is shown in Figure 7.9. In the absence of an applied electric field, incident photons having energy slightly smaller than the bandgap of the intrinsic layer of the EAM, where the light is confined, are not absorbed. When an electric field is applied, the energy bands are tilted, and the photons are absorbed, jumping into the conduction band by tunnel effect.

As the Franz-Keldysh effect, the quantum confined Stark effect exploits the energy band tilt caused by an applied electric field, but it acts on electrons and holes tightly confined in a narrow quantum well, wherein the electrons and holes energy levels are discretized, making the absorption dependence on the applied field different from that of the bulk material. More details can be found in [15]. In an EAM, as in any passive linear medium, the light absorption is usually accompanied by

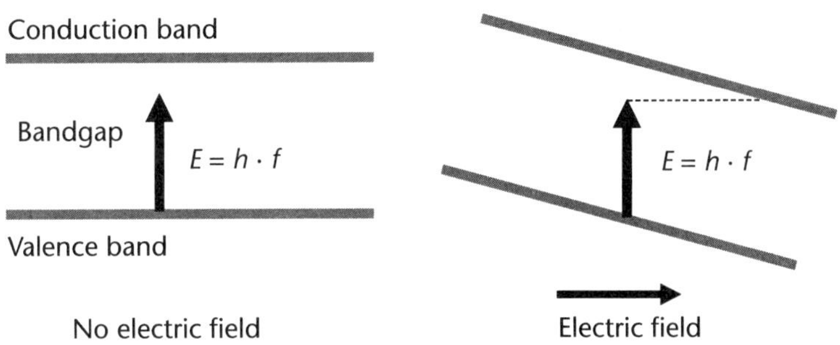

Figure 7.9 Franz-Keldysh effect.

a change of refractive index, due to the Kramers-Kronig relationship between the real part and imaginary part of the propagation constant. The changes of refractive index introduce chirp on the output optical signal, an effect sometimes created on purpose to compensate for the fiber CD. Several design choices are possible for an EAM, such as changing material or geometry, so that it is impossible to give a general law for the output optical power dependence on the modulating voltage. The gain efficiency slope is usually provided by the device manufacturer, together with other data such as optimal bias voltage and chirp coefficient. An example of output power versus voltage curve can be found in [18], together with an analysis of nonlinear distortion introduced by an EAM in an RoF link.

7.5 Design of a Radio over Fiber Link

Design guidelines will be now provided for an RoF link based on the optical technologies discussed in the previous sections. The gain, g, of an RoF link is the ratio between output power and input power, and, with reference to Figure 7.1, it can be calculated as:

$$g = g_{LNA} \cdot g_{OL} \cdot g_{POST} \tag{7.23}$$

an RoF where g_{LNA} and g_{POST} are the gains of low-noise amplifier and the postamplifier, respectively, and g_{OL} is the optical link gain, given by:

$$g_{OL} = \left(\frac{s \cdot \Re}{l} \right)^2 \tag{7.24}$$

In (7.24), s is the transmitter slope efficiency (see Section 7.4), $\Re$ is the receiver sensitivity, and l is the optical link loss including fiber, connectors, attenuators, and so forth. Another important parameter is the noise figure, *NF*, defined as ratio between the noise power, N_{out}, at the output of the RoF link and the power of input thermal noise, amplified by the link gain g. The power of the input noise is $k \cdot T \cdot B$, where k is Boltzmann's constant, T is the absolute temperature, and B is the RoF link bandwidth [19]. It is:

$$NF = \frac{N_{out}}{k \cdot T \cdot B \cdot g} \tag{7.25}$$

The noise figure can be calculated by means of the usual concatenation rule:

$$NF = NF_{LNA} + \frac{NF_{OL} - 1}{g_{LNA}} + \frac{NF_{POST} - 1}{g_{LNA} \cdot g_{OL}} \tag{7.26}$$

where NF_{LNA} and NF_{POST} are the noise figures of the low-noise amplifier and the postamplifier, and N_{OL} is the noise figure of the optical link, which can be calculated as [20]:

$$NF_{OL} = 2 + \frac{1}{g} + R_L \frac{i_{POST}^2 + 2 \cdot q \cdot i_{PD} + i_{RIN}^2}{g \cdot k \cdot T} \tag{7.27}$$

In (7.26), R_L is the load impedance (e.g., 50Ω), i_{POST} is the noise psd of the postamplifier (e.g., from 8 to 12 pA/√Hz), i_{RIN} is the laser relative intensity noise (e.g., –130 dB/Hz), q is the electron charge ($1.60217662 \times 10^{-19}$ A·s), and i_{PD} is the photodetector current:

$$i_{PD} = \frac{i_{bias} \cdot s \cdot \mathfrak{N}}{l} \tag{7.28}$$

where i_{bias} is the photodiode bias current (e.g., 5 mA). If the CSO distortion is negligible, as it is expected in a well-designed RoF link, the link nonlinearity is given by the OIP_3 (Figure 7.6). It can be calculated by concatenating the contributions of the different elements of the link [20]:

$$OIP_3 = \left[\left(OIP_{3,LNA} \cdot g_{OL} \cdot g_{POST} \right)^{-1} + \left(OIP_{3,OL} \cdot g_{POST} \right)^{-1} + \left(OIP_{3,POST} \right)^{-1} \right]^{-1} \tag{7.29}$$

In (7.29), $OIP_{3,LNA}$, $OIP_{3,OL}$ and $OIP_{3,POST}$ are the contributions of the low-noise amplifier, the optical link and the postamplifier, respectively. It is:

$$OIP_{3,OL} = OIP_{3,l} \cdot \left(\frac{\mathfrak{N}}{l} \right)^2 \tag{7.30}$$

where $OIP_{3,OL}$ is the OIP_3 of the DML, or the external modulator. An important parameter of an RoF link is the spurious free dynamic range (SFDR). In an ideal system, no power should be present out of the signal bandwidth, around the radio carrier frequency. In a real nonlinear system, a significant amount of power is detected in correspondence of the harmonics of the radio carrier, called spurs. The SFDR measures the ratio between the amplitude of the radio carrier and the amplitude of the most prominent harmonic, usually the third-order one. A simple equation links the SFDR, measured in dB/Hz$^{2/3}$, to OIP_3, gain and noise figure, all expressed in decibel units [20]:

$$SFDR = \frac{2}{3}\left(174 + OIP_3 - G - NF\right) \tag{7.31}$$

The addend equal to 174 in (7.31) is due to the thermal noise floor, which is −174 dBm/Hz for a 50Ω load at 25°C [21].

7.6 Performance Analysis of a Subcarrier Multiplexing System

Although RoF is considered as an analog transmission system, the modulating signal actually consists of digitally modulated subcarriers, whose performance can be individually analyzed with the methods illustrated in Chapter 4. In this section, an SCM signal will be used as example. Under the approximations discussed at Section 7.2, the current, I_{SCM}, generated by an SCM signal with N subcarriers at the output of a photodiode, can be written as:

$$I_{SCM}(t) = I_0 \cdot \left[1 + m \cdot \sum_{n=1}^{N} u_{I,n}(t)\cos(\omega_n \cdot t + \vartheta_n) + u_{Q,n}(t)\sin(\omega_n \cdot t + \vartheta_n)\right] \tag{7.32}$$

where I_0 is the average photocurrent; m is the modulation index; $u_{I,n}(t)$ and $u_{Q,n}(t)$ are the in-phase and in-quadrature components of the nth subcarrier; ω_n is the frequency (radian/s) of the πnth subcarrier; and ϑ_n is the phase of the nth subcarrier. The subcarrier phases are assumed to be independent and uniformly distributed between 0 and 2π. It will be also assumed that all subcarriers are 16-QAM modulated. According to the methods illustrated in Section 4.4, the following variables can be defined with reference to the 16-QAM constellation of Figure 7.10, having I and Q amplitude levels equal to ±1, and ±3. The dashed lines in Figure 7.10 border the receiver decision zones for each symbol.

The lowest amplitude level is i_S, which equals half the horizontal or vertical distance between adjacent symbols:

$$i_S = m \cdot I_0 \tag{7.33}$$

The receiver SNR is defined as:

$$snr = \frac{i_S}{\sigma_N} \tag{7.34}$$

where σ_N is the standard deviation of the noise, assumed to be equal for all the symbols, and given by (4.46):

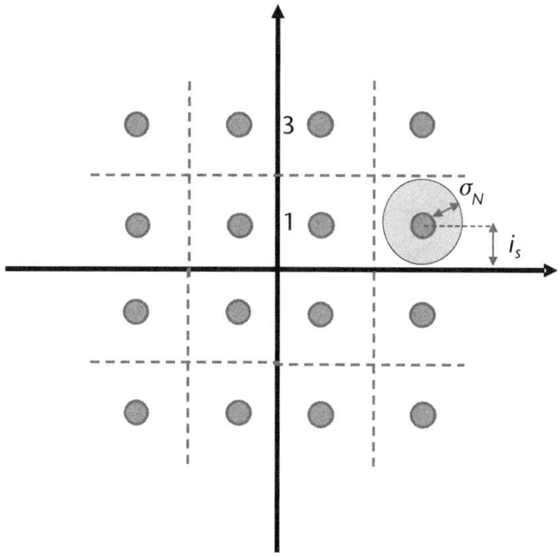

Figure 7.10 16-QAM constellation.

$$\sigma_N = \sqrt{\sigma_{therm}^2 + \sigma_{shot}^2 + \sigma_{RIN}^2} \tag{7.35}$$

σ_{shot}^2 is the shot noise (4.46):

$$\sigma_{shot}^2 = 2 \cdot q \cdot (I_0 + I_d) \cdot B_E \tag{7.36}$$

where q is the electron change, I_d is the dark current, and B_E is the receiver bandwidth. σ_{therm}^2 is the thermal noise (4.43) and σ_{RIN}^2 is the Relative Intensity Noise (4.43) given by:

$$\sigma_{RIN}^2 = rin \cdot I_0^2 \cdot B_E \tag{7.37}$$

Additional noise terms are present with optical amplification, or when the fiber nonlinearity is significant. They can be calculated as illustrated in Sections 4.4 and 4.5, so the analysis will not be repeated here. It is just mentioned that XPM and FWM are especially detrimental in an SCM system, due to the narrow frequency spacing of the subcarriers. Their noise variances are calculated as in Sections 4.5.5 and 4.5.6, replacing wavelengths with subcarriers. The symbol error rate, SER, of the subcarrier depends on the snr and the number of constellation symbols, M (M = 16 for 16-QAM):

$$SER(snr) = \frac{1}{M} \cdot \left[\left(\sqrt{M} - 2\right)^2 \cdot P_1(snr) + 4 \cdot \left(\sqrt{M} - 2\right) \cdot P_2(snr) + 4 \cdot P_3(snr) \right] \tag{7.38}$$

The terms P_1, P_2, and P_3 at the right side of (7.42) are the error probabilities of the four inner symbols, the eight outer symbols, and the four corner symbols of the 16-QAM constellation in Figure 7.10 (indeed, 4 + 8 + 4 = 16). They can be calculated subtracting to 1 the probability that a symbol is correctly decided, integrating the probability density functions of the symbols, assumed to be Gaussian and independent, within decision zones shown in Figure 7.10.

$$\begin{aligned}
P_1(snr) &= 1 - \overbrace{\left[1 - 2\cdot Q(snr)\right]}^{\text{Probability that the I symbol is correctly decided}} \cdot \overbrace{\left[1 - 2\cdot Q(snr)\right]}^{\text{Probability that the Q symbol is correctly decided}} = 4\cdot Q(snr) - 4\cdot Q^2(snr) \\
P_2(snr) &= 1 - \overbrace{\left[1 - Q(snr)\right]}^{\text{Probability that the I symbol is correctly decided}} \cdot \overbrace{\left[1 - 2\cdot Q(snr)\right]}^{\text{Probability that the Q symbol is correctly decided}} = 3\cdot Q(snr) - 2\cdot Q^2(snr) \\
P_3(snr) &= 1 - \overbrace{\left[1 - Q(snr)\right]}^{\text{Probability that the I symbol is correctly decided}} \cdot \overbrace{\left[1 - Q(snr)\right]}^{\text{Probability that the Q symbol is correctly decided}} = 2\cdot Q(snr) - Q^2(snr)
\end{aligned} \tag{7.39}$$

In Equation 7.39, $Q(x)$ is the gaussian Q function

$$Q(x) = \frac{1}{\sqrt{2\pi}} \int_x^{\infty} e^{-\frac{\xi^2}{2}} d\xi \tag{7.40}$$

For 16-QAM:

$$SER(snr) = 3\cdot Q(snr) \tag{7.41}$$

The same result can be obtained calculating the SER of a PAM-4 signal [22] and doubling it to consider both the *I* and *Q* signals [23]. With a Gray code, adjacent symbols differ in only one bit, so that a decision error on one symbol approximately correspond to a single bit error:

$$BER \sim \frac{SER}{log_2(M)} \tag{7.42}$$

It is customary to assess the performance of an RoF system in terms of EVM, defined in Section 4.3.4. For the SCM system under analysis, it can be easily demonstrated that the EVM numerator (4.21) is $\sqrt{2}\cdot\sigma_N$. The constant A at the EVM denominator can be set to be equal to the maximum module among the constellation symbols or to the root mean square value of the symbols. In the first case, it is

$$A = \sqrt{2}\left(\sqrt{M} - 1\right)\cdot i_S \quad \text{and} \quad EVM = \frac{1}{\left(\sqrt{M} - 1\right)\cdot snr} \tag{7.43}$$

In the second case, to calculate the mean square value of a M-QAM constellation, it is observed that it equals the mean square value of the M-PAM constellation on one of the I or Q modulator branches. It can be calculated using the sum of the first N odd numbers:

$$\sum_{n=1}^{N}(2n-1)^2 = \frac{4\cdot N^2-1}{3}\cdot N \tag{7.44}$$

Putting $N = \sqrt{M}/2$, multiplying the result by two, to consider the negative symbols, and, finally, dividing by the number of I or Q symbols, $\sqrt{M}$, it is obtained:

$$A = \sqrt{\frac{M-1}{3}}\cdot i_S \quad \text{and} \quad EVM = \sqrt{\frac{6}{M-1}\frac{1}{snr}} \tag{7.45}$$

Using (7.43) or (7.45), it is possible to calculate SER and BER as function of the EVM rather than the SNR.

The nonlinearity of the optical modulator can have a significant impact on the performance of a SCM system. With an MZM, it can be demonstrated [24] that, for each subcarrier, the mean square value of the photocurrent can be written as the sum of several terms I_k^2, where the subscript k indicates the order of distortion, with $k = 1$ corresponding to the signal. The term I_k^2 takes the form:

$$I_k^2 = 2\cdot I_0^2 \cdot N_k \cdot J_1^{2n}(m)\cdot J_0^{2N-2n}(m) \tag{7.46}$$

where $J_0(\cdot)$ and $J_1(\cdot)$ are order-0 and order-1 Bessel's functions, and N_k is the number of kth order intermodulation products ($N_1 = 1$, $N_3 = 3/8\cdot N^2$, $N_3 = 1/12\cdot N^4$). The carrier-to-intermodulation ratio (CIR) is defined as the ratio between the signal and the sum, σ_I^2, of the distortion terms:

$$CIR = \frac{I_1^2}{I_3^2 + I_5^2 + \ldots} = \frac{I_1^2}{\sigma_I^2} \tag{7.47}$$

In (7.47), the third-order intermodulation is the dominant term and higher-order terms are usually neglected. σ_I^2 can be put as an additional noise contribution in (7.35), after it has been properly normalized so that $I_1 = i_s$. The ratio between the signal and the sum of noise and intermodulation is called carrier to noise ratio (CNR):

$$CNR = \frac{I_1^2}{\sigma_I^2 + \sigma_N^2} \tag{7.48}$$

7.7 Summary

This chapter discussed analog optical transmission in radio over fiber systems but many of the considerations made therein have a more general scope, such as the analysis of nonlinear distortions and PAPR, which also apply to the air interface of a mobile system, and the optical modulation theory, which is valid also for digital communication systems. In the last part of the chapter, the link modeling techniques developed in Chapter 4 were applied to a SCM system with 16-QAM modulation, to provide the reader with a diverse example for analysis techniques that are usually discussed just for OOK modulation.

References

[1] van Nee R., de Wild A, "Reducing the Pak-to-Average Power Ratio of OFDM," *Proc. VTC '98. 48th IEEE Vehicular Technology Conference. Pathway to Global Wireless Revolution,* Ottawa, Ontario, Canada, May 1998.

[2] Ochiai ,H., and H. Imai, "On the Distribution of the Peak-to-Average Power Ratio in OFDM Signals," *Transaction on Communications,* Vol. 49, No. 2, February 2001.

[3] Muller ,S. H., and J. B. Huber, "A Comparison of Peak Power Reduction Schemes for OFDM," *Proc. GLOBECOM 97. IEEE Global Telecommunications Conference*, Phoenix Arizona, November 1997.

[4] Breiling ,H., S. H. Muller-Weinfurtner, and J. B. Huber, "SLM Peak-Power Reduction without Explicit Side Information," *Communications Letters,* Vol. 5, No. 6, June 2001.

[5] Ahn, H.- J., Y. Shin, and S. Im, "A Block Coding Scheme for Peak-to-Average Power Ratio Reduction in An Orthogonal Frequency Division Multiplexing System," *Proc. VTC2000-Spring. 2000 IEEE 51st Vehicular Technology Conference*, Tokyo, Japan, May 2000.

[6] Fan, P., and X.-G Xia, "Block Coded Modulation for the Reduction of the Peak to Average Power Ratio in OFDM Systems," *Proc. WCNC. 1999 IEEE Wireless Communications and Networking Conference*, September 1999.

[7] Davis, J.A., and J. Jedwab, "Peak-to-Mean Power Control in OFDM, Golay Complementary Sequences, and Reed–Muller Codes," *Transactions on Information Theory,* Vol. 45, No. 7, November 1999.

[8] Schmidt, H., and K.- D. Kammeyer, "Reducing the Peak to Average Power Ratio of Multicarrier Signals by Adaptive Subcarrier Selection," *Proc ICUPC '98, IEEE 1998 International Conference on Universal Personal Communications.* Conference proceedings, Florence, Italy, October 1998.

[9] Bjork, J., and Y. Yamamoto, "Analysis of Semiconductor Microcavity Lasers Using Rate Equations," *Journal of Quantum Electronics,* Vol. 27, No. 11, November 1991.

[10] Tomkos, I., et al., "Extraction of Laser Rate Equation Parameters for Representative Simulations of Metropolitan-Area Transmission Systems and Networks," *Optics Communications,* Vol. 194, No. 1–3, July 2001.

[11] Martínez, J. J., et al., "Novel WDM-PON Architecture Based on a Spectrally Efficient IM-FSK Scheme Using DMLs and RSOAs," *Journal of Lightwave Technology,* Vol. 26, No. 3, February 1, 2008.

[12] Lau, K. Y., and A. Yariv, "Intermodulation Distortion in a Directly Modulated Semiconductor Injection Laser," *Applied Physics Letters,* Vol. 45, No. 10, November 1984.

[13] Byung, G. K., et al., "RoF-Based Mobile Fronthaul Networks Implemented by Using DML and EML for 5G Wireless Communication Systems," *Journal of Lightwave Technology,* Vol. 36, No. 14, July 2018.

[14] Yamamoto, T., "High-Speed Directly Modulated Lasers", https://pdfs.semanticscholar.org/bf52/566fcf24d68d9ac01352af1138f7f13ce906.pdf.

[15] Li, G. L., and P. K. L. Yu, "Optical Intensity Modulators for Digital and Analog Applications," *Journal of Lightwave Technology,* Vol. 21, No. 9, September 2003.

[16] Hui, R., et al., "Subcarrier Multiplexing for High-Speed Optical Transmission," *Journal of Lightwave Technology,* Vol.20, No. 3, March 2002.

[17] Abdul-Rashid, H. A., et al., "Chromatic Dispersion Penaltyin Orthogonal Subcarrier-Optical Tandem Single Side Band Systems," *Proc. 2005 13th IEEE International Conference on Networks,* jointly held with the 2005 IEEE 7th Malaysia International Conference on Communication, November 2005.

[18] Sun, C. K., S. A. Pappert, and R. B. Welstand, "High Spurious Free Dynamic Range Fibre Link Using a Semiconductor Electroabsorption Modulator," *Electronics Letters*, Vol. 31, No. 11, May 1995.

[19] Wake, D., A. Nkansah, and N. J. Gomes, "Radio Over Fiber Link Design for Next Generation Wireless Systems," *Journal of Lightwave Technology*, Vol. 28, No. 16, August 2010.

[20] Sauer, M., A. Kobyakov, and J. George, "Radio Over Fiber for Picocellular Network Architectures," *Journal of Lightwave Technology*, Vol. 25, No. 11, November 2007.

[21] Lona, D. G., et al., "Investigation of Noise Sources in Radio-over-Fiber Systems for Wi-Fi Application," *Proc. 2011 SBMO/IEEE MTT-S International Microwave and Optoelectronics Conference (IMOC 2011)*, November 2011.

[22] Carlson, A. B., P. B. Crilly, and J. C. Rutledge, *Communication Systems*, McGraw-Hill, 2002, p. 457.

[23] Proakis, J. G., *Digital Communications*, New York: McGraw-Hill, 1995.

[24] Olshansky, R., "Optimal Design of Subcarrier Multiplexed Lightwave Systems Employing Linearized External Modulators," *Journal of Lightwave Technology*, Vol. 10, No. 3, March 1992.

CHAPTER 8

Photonics for Radio Systems and Networks: Optical Beamforming

8.1 Introduction

To address the challenge of providing extremely high traffic capacity and multi-Gbps data rates in ultradense network deployments, arrayed antennas using the radio beamforming techniques are introduced; to steer the beam in both the vertical and horizontal planes; to reduce the interference between densely deployed cells; and to increase the wireless data throughput. New enabling technologies can be adapted to satisfy current and future stringent requirements in the new generation of ICT networks. Photonics can play the rule overcoming the main technological limitations for radio signal processing at very high frequency allowing for high bit rate transmission. This chapter is organized into several sections: In Sections 8.2 and 8.3, the benefits of beamforming in the evolution of 5G mobile networks are introduced through research and development work. In Sections 8.4, 8.5, 8.6, and 8.7, the concept of beamforming, its requirements, and its conventional implementation are analyzed. In Sections 8.8, 8.9, 8.10, and 8.11 the benefits of using photonics are presented and examples of optical beamforming realizations are described.

8.2 Aspects of the Next Generation ICT that Make Beamforming a Key Block of Such Systems

New types of devices and services related to connected things are an integral part of the most advanced 4G telecommunication systems. Examples of these connected things include connected traffic lights, vehicles, medical devices, waste bins, and electricity supply systems—basically anything that can benefit from a connection. This connectivity brings significant benefits to people, businesses, and society.

In this direction, key challenges can be addressed by 5G and future evolution; massive growth in traffic volume, 1,000 times current traffic; massive growth in connected devices including device-to-device communication in robotics, industries, domotics, and security applications; a wide range of requirements and characteristics in terms of data rates, latency, reliability, and device energy consumption. To this aim, the technology based on beamforming antennas has been introduced in 5G mobile access networks. The beamforming antennas improve the spectrum usage, beyond the LTE frequency spectrum.

Mobile communication systems operating in millimeter wave bands have been seriously considered by technology and system suppliers as the very promising approach to boost capacity significantly [1, 2]. Beamforming antennas are integrated with the 5G system to cope with the main requirements of bandwidth increase with optimized power consumption and footprint. Energy efficiency should also be an important focus area, even with the expected massive increase in traffic reported. By 2024, the 5G technology must be ready to harvest more than 400 exabytes per month of total fixed and mobile data traffic [1]. To hold the huge traffic volume, 5G must be more than ten times faster than 4G in terms of peak speed, offering a download speed of 10 Gb/s and beyond (4G only offers 1 Gb/s). Beamforming antennas are the viable solution and are presented in this work as a step toward realizing the vision of unlimited access to information

Photonics technologies can be the differentiator in beamforming antenna technology evolution. Thus, beamforming antennas equipped with photonics circuitry can be the first examples of how optics can push forward innovation in the next generations of wireless networks, and more in general in the continuously evolving ICT networks.

8.3 Impacts of Beamforming Antennas in New Generation Wireless Networks and Future Scenarios

To increase transmission capacity and achieve good link budgets while reducing interference, smart antennas with beamforming techniques can be designed. The beamforming antenna consists of an array of elements (e.g., 8 × 8 and 16 × 16 elements or more) able to move the beam both in vertical and horizontal plane considering that the main application of such technology will be the urban areas. Beamforming antennas will be applied in the context of micro- and picocells, so cells with a radius of 200m and 10m, respectively. Picocells are thought for indoor application or for very crowded areas.

Considering current sector sizes (see Figures8.1(a) and (b)), supposing to cover each sector with a single antenna, a shift of at least +/– 30 degrees in horizontal direction can be considered. Adding a sufficiently large vertical elevation can realize more directed and spatially separated transmissions to a larger number of users, leading to an additional increase in throughput and coverage (see Figure 8.1(c) and (d)). With beamforming antennas, a significant system performance improvement can be enabled, by serving mobile data traffic with increased capacity (e.g., up to 10 Gb/s) to different final users, while they are moving in the inner and outer LTE cell area or located at different floors of high-rise buildings.

To address the challenge of providing extremely high traffic capacity and multi-Gbps data rates in specific scenarios, the beamforming antennas enable the deployment of ultradense network with nodes operating with very wide transmission bandwidths in higher-frequency bands than installed LTE and relying on new technological solutions.

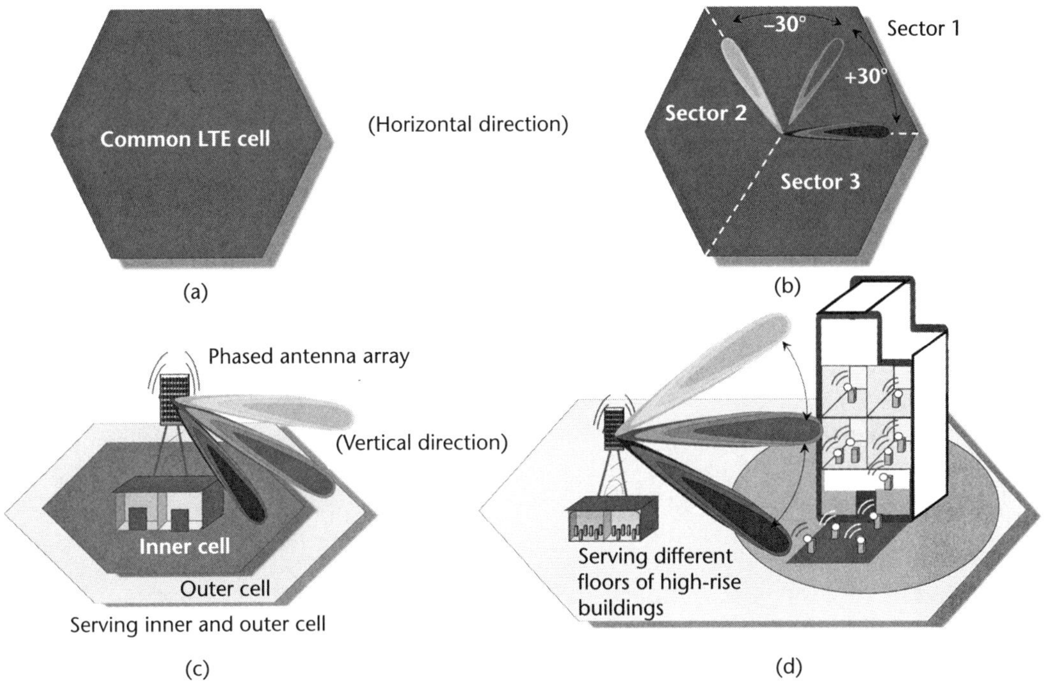

Figure 8.1 (a) The most common cell configuration for LTE, (b) the relevant three-sectored solution, and (c) the beamforming evolution in the vertical direction, and (d) serving different floors of high-rise buildings.

Ultradense networks in 5G applications can be designed with beamforming antennas and low-power access nodes deployed with a higher density than 4G networks with LTE antennas. Access nodes can be located in every room in the case of indoor deployments and at lampposts in the case of outdoor deployments.

To reliably support multi-Gbps data rates, ultradense networks enabled by beamforming antennas should support minimum transmission bandwidths of several hundred megahertz with the possibility of an extension up to a few gigahertz. The new network generation will primarily operate in the millimeter wave range (e.g., 10–100 GHz).

The adopted frequencies in microcells are subject to licensing to get high connection reliability. The unlicensed frequencies around 60 GHz will be mainly used for picocells (indoor application) where the link distance is shorter, and the enhanced directivity of beamforming antennas can be used to avoid interference. 1 Gbps is expected to be provided uniformly within a cell with a peak rate of 5 Gbps for mobile application. In the case of a more nomadic user, higher peak rates can be reached up to 10 Gbps. For industrial communication and societal functions, such as traffic safety, e-health, and smart city management, the reliability requirements also on latency have primary importance and cannot always be met by LTE. Certain use cases involving smart-grid communication applications and traffic safety, for example, require a network latency that is lower than milliseconds end-to-end at any network location.

To meet requirements on low latency, the RTT should be significantly shortened to give the users access to high resource consuming applications like video and augmented reality, and to a continuously increasing number of new services.

Latency for time critical services should be guaranteed with 99.999% reliability. To achieve this goal, the important function of transmission link characterization is performed in beamforming antennas and in other smart antennas like Massive MIMO to optimize the antenna main parameter settings and to precisely find the final user geolocation. A dedicated communication link can be established and the shortest path possible can be found to optimize latency and bandwidth availability with high reliability.

Another important aspect to not exclude for 5G and its developments is the frequency accuracy and jitter affecting the radio signal generation and detection. Already with LTE advanced the requirements for frequency and phase synchronization accuracy are 50 parts per billion, and 50 parts per billion per milliseconds, respectively (using IEEE 1588 Precision Time Protocol). For 5G, and thanks to the technological advances driving the beamforming antenna evolution, it can be expected that such values will be strongly improved. In 5G evolution, time synchronization can be expected to be lower than tens of nanosecond.

8.3.1 The Innovation Impact on System Complexity and Footprint

However, the innovation based on smart antennas for introducing beamforming could imply an increase in terms of power consumption, complexity and cost. Since the deployment of 5G will imply a huge number of base stations and array antennas, it becomes important to push towards an extreme integration of devices to reduce their footprint, making the solution feasible. At the same time, the power consumption should be reduced to make possible deployment of the high number of required base stations with a reasonable value of energy consumption. As a starting point, it can be assumed that power consumption should be kept at the current level, increasing the traffic volume 1,000 times. In this case, the introduction of new technological advances can help in implementing a very compact, low cost, and low power consuming beamforming solution with outstanding radio performances (e.g., very low phase noise and very limited squint phenomenon).

8.4 How to Operate with Beamforming Antennas

Beamforming antennas are introduced in 5G systems, to have the capability of pointing the radio beam to cover a precise area, tuning its orientation to reduce interference with closed radio sectors in such a way that it is possible to cover several small crowded areas with radio beams working at millimeter and microwave frequencies.

In perspective, it is thought to potentiate beamforming, introducing the capability of following users with the radio beams, having multiple beams irradiated by the same antenna. However, to cover the same area or close area with radio beams, the latter must work using different frequencies to avoid interference, so several licensed frequencies must be adopted. Phased-array antennas are widely used in radar systems and are also being introduced into mobile communications networks using pico and nanocells to serve densely populated areas with high speed and high quality of service communications links. Due to their reduced size, pico- and nanocells face an increased possibility of interference between adjacent cells;

beamforming using phased-array antennas is being considered to control the directivity of the radio frequency signal beam, to minimize interference. Phased-array antennas for 5G applications have millimeters-square dimensions and can consist of an array of hundreds of phase-controlled radiation elements. Figure 8.2 encompasses 64 similar radiating elements which can have different shapes and design [3].

Phase antenna arrays are characterized by having a radiation lobe that is assembled from many similar radiating antenna elements, such as slots or dipoles, each of which transmits or receive a radio signal having an individually controlled phase and amplitude (see Figure 8.3(a)). The phase can be determined by calculating the delay experienced by the signal in arrival to each phased-array antenna element. For instance, when the elements are displayed as in Figure 8.3(a) and assuming that c is the light speed in free-space, the delay is:

$$\Delta t = (d \sin\theta)/c \tag{8.1}$$

The phase shift at a given frequency can be obtained by:

$$\Delta\varphi = 2\pi f \ \Delta t = (2\pi f \ d \sin\theta)/c \tag{8.2}$$

Accurately predictable radiation patterns and beam-pointing directions are achieved with the beams to have high gain and low sidelobes, or controlled beam width. Beamforming modulates the signals emitted by an array of antennas both in amplitude and phase to direct the propagation of the radio signal in the desired direction. The technology proposed allows the manipulation of the radio waves using proper antenna surface realizations; an example is reported in Figure 8.2.

The standard solutions are based on an alternate phase and amplitude control into a structure, which provides the RF power distribution across the aperture area of the antenna array. This area can be as simple as a planar surface, which is illuminated by an RF signal and provides a known reference near-field distribution over the aperture. The goal is to then find a modulation function whose interaction transforms the reference wave into the desired aperture distribution. All solutions

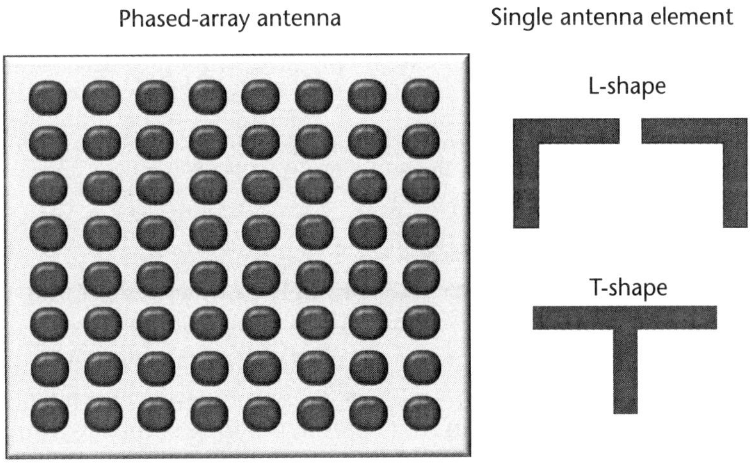

Figure 8.2 Typical scheme of a phased-array-antenna with examples of radiating elements

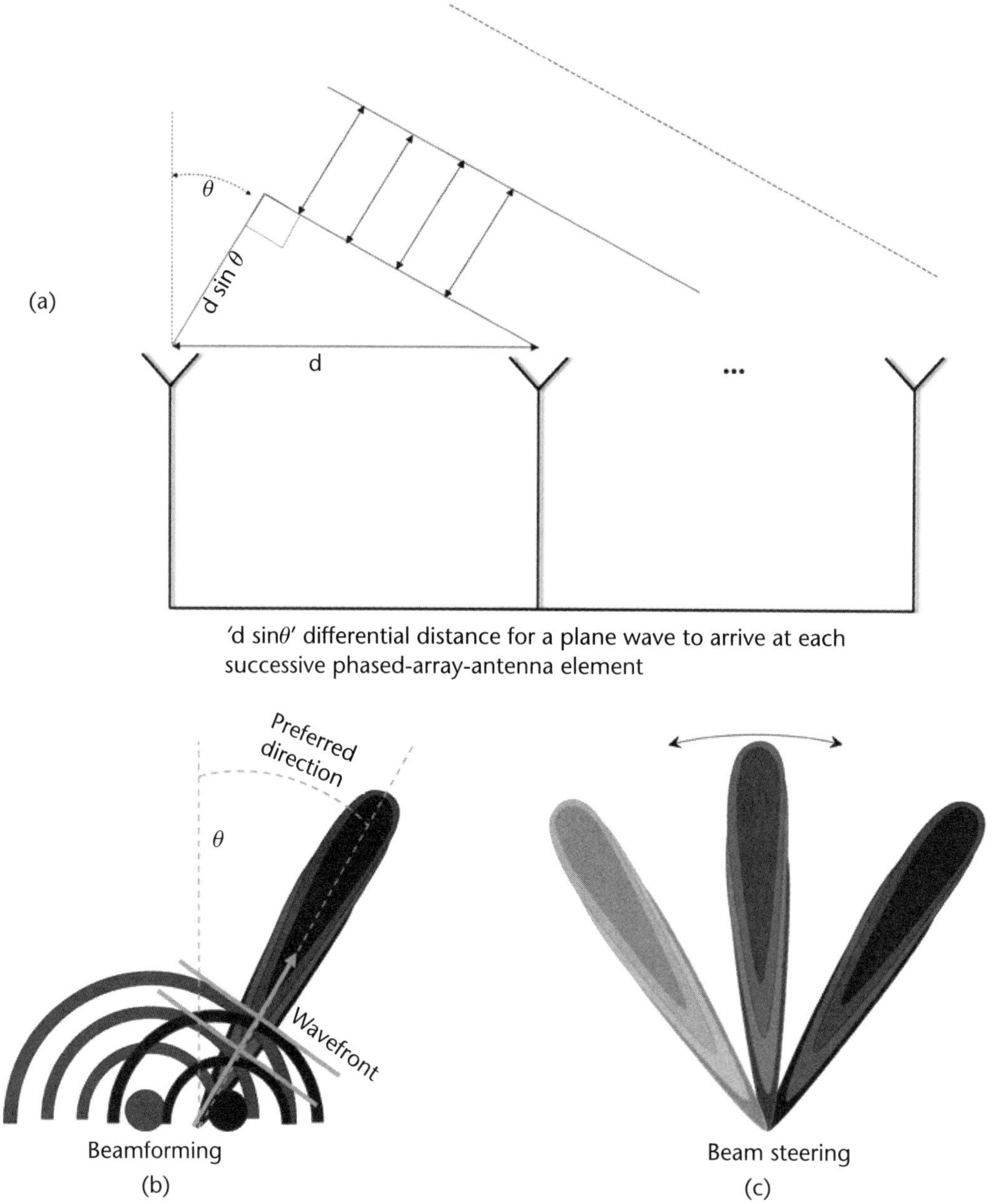

Figure 8.3 Schematic of an array of radiating elements: (a) differential distance, (b) beamforming, and (c) beamsteering performed by wave-front rotation with phase-controlled elements.

require a biased control component, like a varactor or transistor, at each emitting dipole element, but being driven by a single baseband unit radio control.

The design of an antenna enabling the above modulation function can often be nontrivial. The modulation function is an algorithm to generate in the near field a wanted field distribution (e.g., typically Sinc^2 versus azimuth angle, θ) that is used to obtain in the far field a tuneable pencil, pointing in the horizontal and azimuthal direction with efficient antenna gain, as in Figure 8.3(b) and(c).

8.5 What a Beamforming Antenna Looks Like

The beamforming is achieved by controlling the mutual phase relationship among the wave-fronts in the near field, generated by the different RF-emitting elements of the antenna array so that all signals are in phase only in correspondence of the desired direction. The beam steering is achieved by properly changing the mutual phase relations to obtain in the far field a pencil pointing in a tuneable horizontal and azimuthal direction with an efficient antenna gain for each used frequency.

Each radio base station equipped with a beamforming antenna transmits an M-QAM signal, whose components are the in-phase (x_I) and quadrature (x_Q) components of the RF signal of information toward the final user. The main difference respect to a usual M-QAM RF transmitting antenna is given by the presence of a phase modulator per each antenna element that modulates the RF carrier with a specific phase variation to steer a beam in a wanted direction. Suppose having N antenna elements, each fed by a single RF carrier $f = \omega/(2\pi)$; each user will receive a signal $y_m(t)$ given by

$$y_m(t) = \sum_{n=1}^{N} \left(x_{Im} \cos(\omega_m t + \varphi_n(t)) + x_{Qm} \sin(\omega_m t + \varphi_n(t)) \right) \tag{8.3}$$

where $\varphi_n(t)$ represents the phase, given by the modulation function, and x_{Im} and x_{Qm} are the in-phase and quadrature components of the information symbol sent to the m user on a carrier ω_m at time t. The modulation function $\varphi_n(t)$ can be constant in case the final user is static, or can vary versus time to perform target tracking in case the final user is traveling on an unpredictable path in proximity, or far from the antenna, in a metropolitan or rural area, and/or with different link budgets.

Given an array antenna structure with N elements, all capable of transmitting at the same time on different carriers toward different users, the modulation function can generate a different beam steering for each user on different frequencies, and the overall transmitting function $Y(t)$ can be in the following formula:

$$Y(t) = \sum_{m} a_m \sum_{n=1}^{N} \left(x_{Im} \cos(\omega_m t + \varphi_n(t)) + x_{Qm} \sin(\omega_m t + \varphi_n(t)) \right) \tag{8.4}$$

where a_m is an adaptive real coefficient depending on the Euclidean distance for the m user relating to the radio base station in the cell. The transmission link performed by the beamforming antennas is bidirectional; downlink and uplink can be performed by the same antenna element. The communication from the radio base station to the final user is the downlink, and vice versa the uplink is defined. In the bidirectional link, two main different antenna schemes can be designed; the TDD and the FDD arrangement.

When the RF signal is sent to the antenna element for transmission, the receiver should be isolated to avoid that an echo of the transmitted signal is coupled with the rest of the circuitry. The TDD solution utilizes a fast circuit switch to connect the antenna to the transmitter in transmission mode and to the receiver otherwise. To optimize bandwidth occupancy the uplink and the downlink can be performed in alternate and equal time intervals.

The FDD performs circuit isolation with adequate passband filters by transmitting and receiving on two different frequencies, practically filtering out the echoes from the transmitter side at the receiver frontend. The two possible schemes for beamforming transmission with TDD and FDD techniques are represented in Figure 8.4, also outlining the needed circuitry to perform phase and frequency generation on the several antenna elements. The antenna array can be realized with 100 elements to perform beamforming and beam steering of transmitted signals.

8.6 How to Drive Beamforming Antennas

The use of high carrier frequencies in the millimeter-wave range unavoidably requires the implementation of large antenna arrays to attain the same effective antenna aperture of systems operating at carrier frequencies below 6 GHz already in use in LTE. There are substantially three approaches used to drive beamforming antenna; analogic, digital, and hybrid solution.

In the analogic solution, each single antenna element in downlink and uplink (i.e., the transmitted/received signal link at radio base station) is driven in terms of phase and amplitude in RF band by a dedicated analogic circuitry; each one

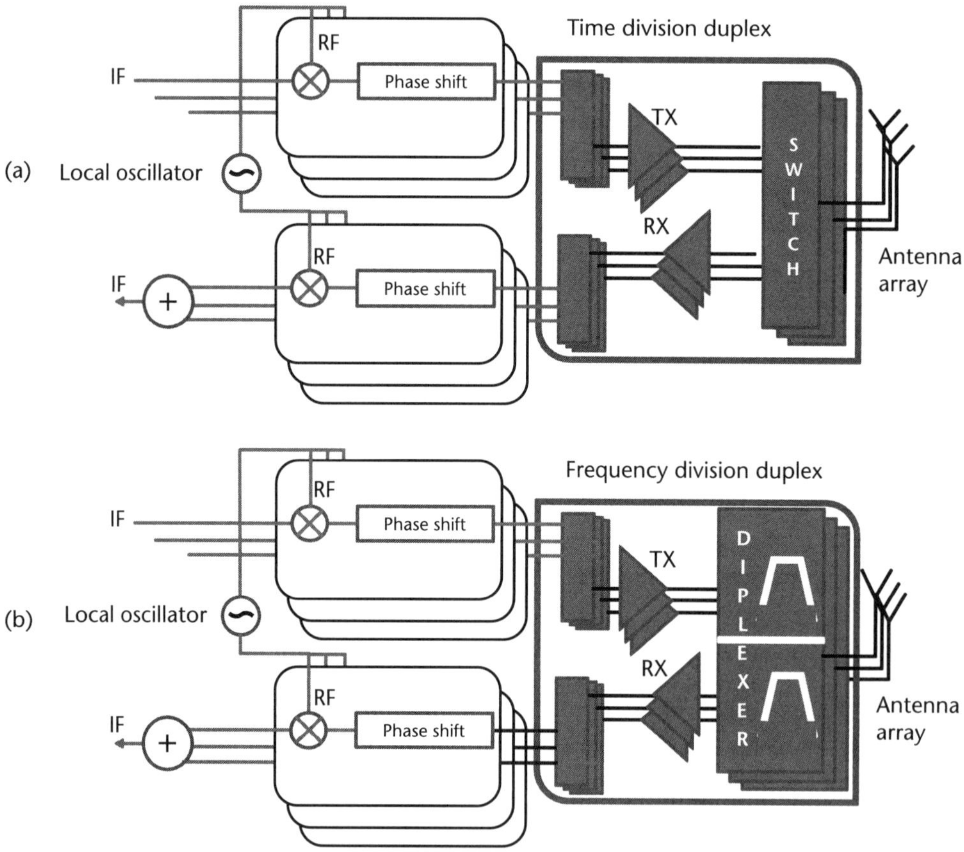

Figure 8.4 Analogic beamforming antenna: (a) TDD, and (b) FDD scheme and phase shifters.

equipped with an analogic phase shifter and amplification stage directly connected to the antenna array. The proper phase shift and electronic gain of each RF signal can be seamlessly changed both at the receiver and the transmitter. At the receiver, the thusly shaped and phase shifted signals coming from each antenna element are summed up in a single analogic signal which is then processed digitally in the baseband unit of the Radio Base Station. The scheme is based on several parallel analogic circuitry which equals the number of radiating elements in the phased-array-antenna. The greater the number of antenna elements, the higher is the complexity of the whole radio system.

The same reflections are valid for the transmitter side. However, the analogic phase shifters and radio frequency amplifiers can perform the beamforming and the beam steering seamlessly with high precision, mainly depending on the phase shifters insertion loss, the maximum electrical RF gain achievable with optimized power consumption, and the number of antenna elements on the antenna surface.

In both cases, the complexity of the scheme can be justified by the need to cover the relevant cells with 99.999% of reliability. In the digital solution, the phase and the amplitude of all RF signals are processed digitally in the digital baseband. Once received, the signals are processed by an ADC circuitry, which is capable of processing in parallel all the data from the different antenna elements, which is then regenerated digitally and summed up in the digital baseband. In the digital approach, parallel functions can be performed, too. For instance, the received information can be regenerated and recovered with high accuracy and the channel quality estimation can be captured at the same time.

Characterization of the communication channel is needed to properly define the functional beamforming shape and direction while performing the connection link. The calculated beamforming parameters can be used at transmitter side to properly drive the phased antenna array, and thus are converted from DAC data to properly modulate the RF transmitted signals, and to optimize the beamforming and the beam steering. An example of a DAC operative in the RF band (e.g., 24 GHz) is reported in Figure 8.5.

Any transmission channel estimation error increases the received noise requiring an extra overhead in the dataflow to properly recover information. The greater the number of ADC digits, the greater the accuracy to estimate the communication link. The sharper the beamforming, the greater the margins of the functional system and the transmitted capacity.

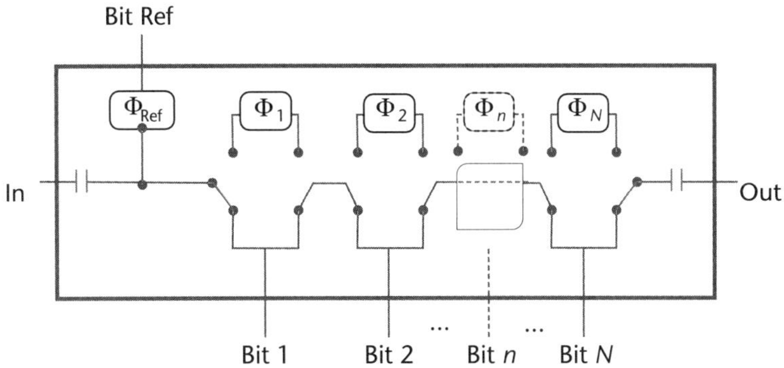

Figure 8.5 Phase shifter N-bit digital phase shifter operative at RF.

There will be another fundamental requirement in tomorrow's radio systems. For example, in the cognitive radio paradigm, an intelligent radio technology will exploit wideband receivers to automatically detect available channels in a wireless spectrum, and constantly monitor the link performance. By adaptively changing the transmission parameters, this technology paradigm will allow for enhancing the performance of the whole wireless communications.

It is noteworthy that to achieve a functional system, both digital and analogic solutions require an upconversion from the baseband to the RF spectrum with high reliability. High-quality oscillators for time and frequency precision are necessary to lower the risk to degrade the transmitted signal and introduce imprecision in the beamforming and the beam steering. To enhance precision in time, frequency, and phase setting for beamforming while being power-efficient, the baseband and radio antenna units' capabilities in the advanced radio base stations must be drastically changed from state-of-the-art antennas for LTE applications.

8.6.1 Phase Precision

The technological advances to enhance system performance require the use of the most advanced technologies to realize the digital units and highly stable clocks. DAC and ADC with a high standard number of effective bits of resolution (e.g., 6–12) and sampling frequency (5–100 GHz) are necessary, given that the local oscillator must perform better than LTE advanced requirements for accuracy (IEEE 1588).

The discretized values of the phase and amplitude values, dictated by the limited resolution of ADC and DAC, or the use of analog phase shifter with limited accuracy, introduce some imprecision in the phase and amplitude setting for each antenna element, causing distortions in the signal beamforming and beam steering. For instance, when discretizing in the DAC operating in the RF band with N possible digits, the possible phase shifts can be the sum of n different angles, with $n \leq N$.

Noteworthy, if Φ_n is one of the possible discretized angles, the following equation must be verified:

$$\Phi_1 + \Phi_2 + \ldots + \Phi_N + \Phi_{\text{Ref}} = 360° \quad (8.5)$$

following the example reported in Figure 8.5, where

$$\Phi_N = \text{Integer}\left[180/2^{N-n}\right] \quad (8.6)$$

The total phase shift is Φ_{m}, where $m \leq 2^N$ and can be calculated by

$$\varphi_m = \sum b_n \Phi_n + \Phi_{\text{Ref}} \quad (8.7)$$

where b_n can be 0 or 1 depending on which discretized angle is to be summed up to achieve the needed phase shift (see Table 8.1 with the bias truth logic in case of N=6). For instance, $\Phi_1 = \varphi_1$ with $b_1 = 1$ and $b_n = 0$ for every $n \neq 1$.

The phase resolution can be measured with the difference between two consecutive phase shift values

Table 8.1 Bias Truth Logic in Case the Digits are $N = 6$

	Φ_n	Φ_1	Φ_2	Φ_3	Φ_4	Φ_5	Φ_6	Φ_{Ref}		Φ_n	Φ_1	Φ_2	Φ_3	Φ_4	Φ_5	Φ_6	Φ_{Ref}
Φ_n	*Phase shift angle*	5°	11°	22°	45°	90°	180°	Φ_{Ref}	Φ_n	*Phase shift angle*	5°	11°	22°	45°	90°	180°	Φ_{Ref}
Φ_1	5°	1	0	0	0	0	0	1	Φ_{49}	275°	1	0	0	0	1	1	1
Φ_2	11°	0	1	0	0	0	0	1	Φ_{50}	281°	0	1	0	0	1	1	1
Φ_3	16°	1	1	0	0	0	0	1	Φ_{51}	286°	1	1	0	0	1	1	1
Φ_4	22°	0	0	1	0	0	0	1	Φ_{52}	292°	0	0	1	0	1	1	1
Φ_5	27°	1	0	1	0	0	0	1	Φ_{53}	297°	1	0	1	0	1	1	1
Φ_6	33°	0	1	1	0	0	0	1	Φ_{54}	303°	0	1	1	0	1	1	1
Φ_7	38°	1	1	1	0	0	0	1	Φ_{55}	308°	1	1	1	0	1	1	1
Φ_8	45°	0	0	0	1	0	0	1	Φ_{56}	315°	0	0	0	1	1	1	1
Φ_9	50°	1	0	0	1	0	0	1	Φ_{57}	320°	1	0	0	1	1	1	1
Φ_{10}	56°	0	1	0	1	0	0	1	Φ_{58}	326°	0	1	0	1	1	1	1
Φ_{11}	61°	1	1	0	1	0	0	1	Φ_{59}	331°	1	1	0	1	1	1	1
Φ_{12}	67°	0	0	1	1	0	0	1	Φ_{60}	337°	0	0	1	1	1	1	1
Φ_{13}	72°	1	0	1	1	0	0	1	Φ_{61}	342°	1	0	1	1	1	1	1
Φ_{14}	78°	0	1	1	1	0	0	1	Φ_{62}	348°	0	1	1	1	1	1	1
Φ_{15}	83°	1	1	1	1	0	0	1	Φ_{63}	353°	1	1	1	1	1	1	1
Φ_{16}	90°	0	0	0	0	1	0	1	Φ_{64}	360°	0	0	0	0	0	0	1

$$\Delta\varphi = \varphi_m - \varphi_{m-1} \tag{8.8}$$

However, in case of angle discretization, not all contiguous phase shifts have the same difference, and this may cause errors in the evaluation of the beamforming and the beam steering. For instance, if $N = 6$ digits the lowest phase shift can be around six degrees (i.e., $360°/2^6$). Following the example of angle discretization in Table 8.1, where $N = 6$, two consecutive phase shifts, φ_m and φ_{m-1}, can differ by five degrees if m is odd and by six if m is an even discrete number. This is true if the phase shift φ_m is not a multiple-integer value of 45°. In this case, the difference is seven degrees. Hence, depending on the phase shift applied, a different resolution may be experienced with a consequent uncertainty in the beamforming shape and beam steering direction.

8.6.2 Time and Frequency Precision

The high bit rates of the 5G scenario will require RF carriers with improved phase stability, which is a tough task for RF local oscillators (LO), particularly for frequencies above 6 GHz. A fundamental design parameter is the stability of the clock reference to properly set the beamforming in a predetermined instant, both at the transmitter and the receiver side. The limited clock stability in frequency and time may cause the squint phenomenon (see Section 8.7) causing the error of the beamforming alignment angle toward a final target with a consequent decrease of the final user connection quality and link capacity.

To further reduce errors in beamforming and beam steering realization, the local oscillator needs to operate an upconversion to form the digital baseband to the radio frequency domain, which should be ultrastable with the lowest possible noise generation (e.g., mainly phase noise and frequency instability). For instance, for a RF voltage controlled oscillator (VCO) at 15 GHz, the phase noise should be conveniently lower than -100 dBc/Hz at 10 kHz and -105 dBc/Hz at 100 kHz to allow for beamforming and beam steering with a few degrees angle precision (e.g., lower than five degrees) over a 100m link distance [4].

The local oscillator frequency should be locked with the transmission frequency through high-resolution electronic control. Beyond 6 GHz, carrier generation with high stability is challenging and a good tradeoff among the possible solutions can be to generate an intermediate frequency (IF) with a very stable electronic oscillator (e.g., several GHz) and then multiply it to generate the carrier at radiofrequency. With the proposed solution, the precision stability can be achieved, but noise generation and the needed power consumption by the multiplying stage can make the final antenna design complex and not energy efficient to achieve high performance. Substantially, a very accurate system design is crucial.

8.6.3 Hybrid Realization

To partially counteract the effects of possible imprecisions in the performance of both analog and digital circuitry for phase shifting, signal shaping, and carrier and time reference generation, an alternative approach can be defined.

The hybrid realization of beamforming antennas aims to simplify the overall system design by preferring digital solutions when the channel estimation must be performed on simplified antenna design and using the best performing analog components to operate directly in the radio frequency spectrum, instead of at the digital baseband, to enhance precision and accuracy in beamformed signal transmission.

In this approach, some beamforming components like the phase shifters, clocks, and local oscillators can be realized in the best technology possible to achieve high precision limiting the use of the phase discretization (e.g., needed at high frequency to operate with ADC), and to avoid the undesired squint phenomenon (see Section 8.7).

8.7 The Phase Shifter, the Squint Phenomenon, and the True-Time-Delay Technique

The desired phase shift can be applied using phase shifters, which produce a determined phase shift, or time delays, depending on the transmitted frequency. In (8.2), a given frequency in the beam corresponds to a different phase shift. Furthermore, with the discretized phase shift technique, the RF modulated signal is shifted in phase so that wave front interference can generate a directional beam in predefined directions. Practical hardware constraints in discretized phase shifters can induce different time delays for different frequencies in the band of the signal to be transmitted in a precise direction.

On the contrary, a phased-array antenna should have the same phase shift for all frequencies within the band of the signal. Unfortunately, practical implementations with phase shifters are relatively good for narrowband signals and fail when large bandwidth signals must be transmitted with the angle of transmission or reception away from broadside. The overall effect is the squint phenomenon, in which the beam (steered by the in-band frequencies but the carrier) deviates in direction (i.e., squints) as a function of frequency. The broader the signal bandwidth, the greater is the antenna-gain variation over frequency for a given transmission or reception angle. To compensate for the squint phenomenon, in case discretized phase shifters are used, a channel capacity constraint should be imposed. Approaches to eliminate beam squint have been studied, including the implementation of the true-time-delay technique.

True-time-delay consists of generating and seamlessly controlling time-shifted modulated RF signals, one for each element of the phased-array antenna, to enhance the precision on beamforming. The technique is based on the use of elements able to introduce multiple progressive time delays in the radio frequency generator. By introducing a time-delay technique on behalf of the phase shifter, using independent steering of multiple RF signals with a continuous variation of the angle of transmission or reception is possible. The main true-time delay benefit is the reduction of the beam steering uncertainty due to the uniform time delay distribution over the entire signal band, and hence wide bandwidths can be accommodated over time without beam squint.

However, a drawback of the true-time delay technique is the relevant, more complex, design than the phase shifter solutions that is traditionally preferred for steering a phased-array antenna. The unpractical designs presented so far in literature can become feasible if photonics is used for two practical reasons: Photonics is immune to electromagnetic interference, and is natively a broadband technology. By applying silicon photonics technology, practical realization can be implemented; some interesting approaches are reported in the following sections.

8.8 How Optics Can Be Beneficial to Perform Phase Shift Control

Today, the electronic solution for the implementation of a beamforming antenna applies digital phase shift at each antenna element. This solution shows significant insertion loss and power consumption, with a nonnegligible phase error (e.g., few degrees in the horizontal and vertical axis).

An increase in the number of antenna elements in the array is to be expected in 5G systems, and the above limits will progressively affect the running efficiency of the beamforming system. Beamforming systems based on photonics have been proposed, either exploiting discretized phase shifter or the so-called true-time delay.

8.8.1 Optical Phase Shift Realization

Optical beamforming based on discretized phase shift is realized by shifting the carrier with respect to the sidebands. This can be done, for example, by exploiting 2-D arrays of pixels based on liquid crystal on silicon (LCoS) and multiwavelength optical sources [5]. This solution requires that the carrier and its sideband are separated more than the resolution of the pixels, hence it cannot be applicable in all possible system designs. Through the true-time-delay technique, the phase can be continuously changed in the 0°–360° range.

In optics, the RF carrier for the antennas can be generated from the beating of a couple of light beams controlled in phase. For this purpose, a two-wavelength-beam source can be designed. For instance, a two-wavelengths mode-locked laser can be adopted for this purpose, or any other technique able to generate two different wavelengths phase-locked with a precise delta frequency [4].

The solutions based on true-time delay have the advantage of avoiding the squint of the beam, even for broadband signals. Photonics has been proposed for this application, thanks to the broad bandwidth transmission enabled (see Section 8.10). The true-time delay has been proposed by means of chirped fiber Bragg gratings and tunable laser sources [6]. This solution requires one tuneable laser per antenna element; therefore, it becomes impractical if the antenna array has several elements. Another technical implementation of true-time delay consists of realizing a set of different time delays by means of optical switches setting different optical paths for each antenna element [7]. This approach lacks the continuity of the delay that would be necessary for precise beam control.

Alternatively, the RF signal generation and steering can exploit the true-time-delay induced CD to the different modes of a mode-locked laser [4]. This solution also needs RF filters at the antenna elements to select the correct beating after the optoelectrical conversion through a photodiode. Besides the fact that this presents additional complexity, it also implies a reduced efficiency in the RF signal generation process.

To the current state of the art, optical true-time-delay approaches are not appealing for mobile wireless communication, due to combinations of high implementation cost, significant insertion loss, large size, or excessive power consumption. Even in this case, silicon photonics technology can enable feasible and practical implementations of phased-array antennas with enhanced performance.

The following section is an interesting example of true-time-delay application, scalable with the number of antenna elements and using basic optical elements, which can be fully integrated into silicon photonics die. The simple schematic allows for the footprint reduction of the needed optical processing unit with a consequent decrease of the insertion loss of the optical circuitry and an increment of energy efficiency.

8.9 A Viable Optical Beamforming Implementation with True-Time Delay

A possible implementation scheme for beamforming with phase shift or true-time delay is reported in Figures 8.6 and 8.7 for transmitter and receiver respectively. Two outgoing light beams generated from the same reference light source are effectively locked. The Δf between the two light beams corresponds to the radio signal frequency. The two beams are separated using optical selective filters (e.g., an AWG). In the case of the transmitter front end, one beam is modulated to transport radio channels at IF. In the receiver case, one beam is modulated at RF+IF by the signal from a single antenna element. The other optical beam (i.e., on λ_2 in Figure 8.6 and Figure 8.7) is sent to an optical microring resonator, to shift the optical carrier in time, exploiting its CD characteristics. A single optical microring resonator is designed, controlled, and tuned to generate a programmable time shift that

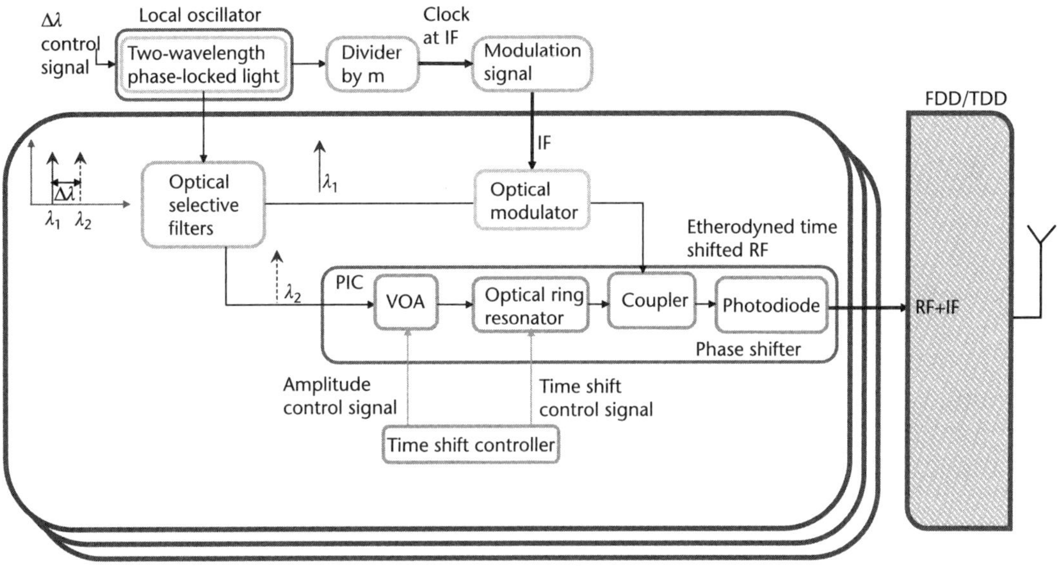

Figure 8.6 Transmitter scheme of true-time-delay beamforming on a single antenna element.

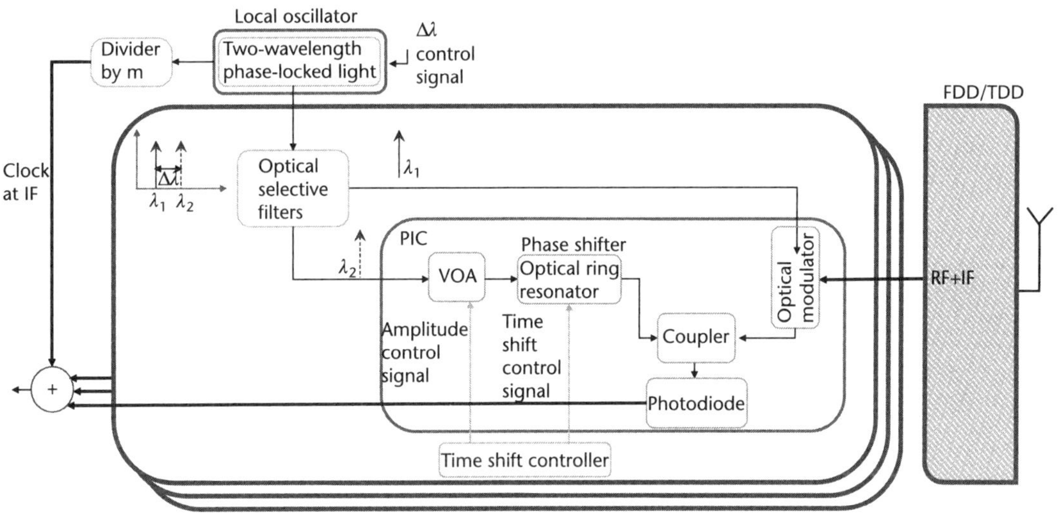

Figure 8.7 Receiver scheme of true-time-delay beamforming on a single antenna element.

corresponds to the time shift required to drive a single element of the beamforming antenna.

The wavelength of the time-shifted optical beam can fall close to the resonance frequency in the aim to achieve an increased value of group delay, without the need to add ring resonators to cumulate further CD. It is worth noting that this way of working cannot be performed in a different schematic as the modulating signal can be filtered by the notch response of the microring resonator at the resonating frequency. The extra losses due to the possible positioning of the time-shifted carrier close to the resonance of the ring resonator can be easily recovered in the scheme, thanks to the resulting beating signal from the optical coupling of modulated and shifted carrier signals; heterodyne detection is performed.

A variable optical attenuator (VOA) is placed in front of the microring resonator to properly compensate the amplitude fluctuations due to the filtering effect of the microring resonator. Once recombined, the two carriers are sent to a photodiode to be directly detected. The intensity of the modulated carrier should be set to compensate the time shift carrier attenuation producing a photocurrent (i.e., resulting from the detection of the beating signal) with enough electrical levels to properly feed the following electronic circuitry. The fundamental aspect is the intensity setting for the modulated signal to generate a proper beating signal, which enables the counteraction of the extra losses experienced by the time shift carrier (a notch response can be measured at the resonance frequency of the ring resonator).

The evaluation above can be derived by considering the basics of the following formula representing, in terms of propagating electrical field in the photonic circuitry, the result of the beating between the time shift beam and the modulated beam.

If the modulated and not modulated signals are represented as

$$S(t) = E_s(t)\cos(\omega_s t + \varphi_s) \tag{8.9}$$

$$P(t) = E_p(t)\cos(\omega_p t + \varphi_p) \tag{8.10}$$

with $E_s(t)$ representing the modulating signal of the form $E_s(t) = a \cdot x(t)\cos(\omega_{IF})$, and $x(t)$ is the IF modulating signal and a is the modulation amplitude. Regulating the amplitude a of the modulation, it is possible to compensate the amplitude reduction of the unmodulated signal, when passing through the VOA and ring resonator chain, to keep the signal level constant.

The current I_{TX} measurable at the photodetector on the transmitter front end, with the bandwidth to detect the RF+IF signal and filter out all other higher frequency components, can be expressed as

$$I \propto A \cdot E_s(t)E_p(t)\cos\left[(\omega_s - \omega_p)\cdot t + \varphi\right] \tag{8.11}$$

The current I_{RX} measurable at the photodetector on the receiver front end, with the bandwidth to detect the IF signal and filter out all other higher frequency components, can be expressed as

$$I \propto A \cdot E_p(t)\cdot a \cdot x(t)\cdot \cos(\omega_{IF}\cdot t + \varphi) \tag{8.12}$$

with $\Phi = (\omega_s\text{-}\omega_p)\,\Delta t$, where Δt is the time shift introduced by the ring resonator. The CD of the optical ring resonator follows a specific profile that is a function of the voltage signal driving it through carrier injection. This characteristic is not linear. A proper control signal is generated as a function of the Δt required.

The attenuation of the ring resonator changes when moving from one Δt value to another with precise behavior. A voltage signal is used then to drive in a coordinated manner the VOA to equalize the total delay to a predefined value. This choice allows for a stable amplitude of the electrical signal generated at the output of the photodiode.

The proposed scheme allows real-time handling of time shift, permitting also the tracking of moving targets with narrow radio beams (i.e., narrow radio lobes). This allows the reduction of power emission and interference with other channels.

Fundamental is the channel characteristics estimation that mainly results in the setting of reference parameters useful for the amplitude control signal and time shift control signal.

This control can be implemented, for example, by means of a common electronic phased-locked loop (PLL).

8.10 The Need for Optical Phase Shift Control

Optics and photonics have the major advantage with electronics counterpart of processing signals at very high frequency natively with high quality and precision. Radar systems and mobile communication network transmitters using phased-array-antennas can adopt an optical radio frequency generation and phase shifters that enable high signal quality operations to be performed on a very stable RF signal. Optical generation of RF signals can be based on heterodyning a pair of phase-locked optical signals that have an optical frequency difference equal to the desired RF signal frequency.

The direction of the resulting beam is controlled by appropriately setting the phase of the RF signal transmitted by each element. The phase is controlled by introducing an appropriate time delay between the optical signals before they are heterodyned.

In one approach, a pair of time-delayed optical signals are generated by subjecting the optical spectrum generated by a mode-locking laser, MLL, to chromatic dispersion by transmitting it through an optical waveguide or fiber. Due to their different wavelengths, the various spectral components in the MLL spectrum experience different delays. By selecting a suitable pair of optical components in the spectrum a specific delay can be introduced, giving the resulting RF signal a specific phase. In an example, a pair of spectral components are selected using a wavelength selective device, like a wavelength selective demultiplexer and switched on a fiber introducing the needed amount of time delay. The used optical components for this approach are typically big and expensive [4–8]. Since a phased-array antenna includes hundreds of elements, it easy to see that such an approach becomes prohibitive in terms of cost and space. The use of an optical fiber to introduce the time delay by exploiting the CD profile of the fiber faces limitations on the total delta delay which can be applied and the delay resolution.

A longer delta delay can be achieved by increasing the CD, whereas a higher resolution is achieved by reducing it.

In another approach, reported in [9], the time delay is introduced by exploiting microring resonator elements (see Figure 8.8) instead of optical fibers to reduce the space occupancy and power consumption. Microring resonators introduce cumulated CD, to generate a group delay in the passing through light. However, this solution faces functional limitations due to the filtering response of the ring resonator, centered at the resonance frequency of the microrings.

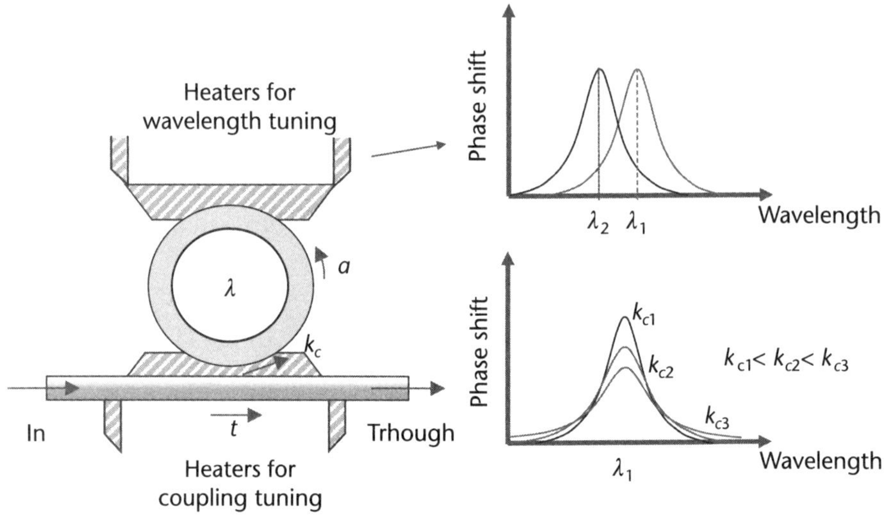

Figure 8.8 Example of group delay profile setting with a single ring resonator designing the ring dimensions and the proper coupling coefficient [9].

To operate with a low optical loss, the operating wavelength needs to be far from the resonance frequency, which unfortunately is in the part of the ring resonator transmission spectrum where the group delay generated by the ring resonator is substantially lower. Therefore, cascaded ring resonators are used to obtain a higher value of group delay, thus complicating the design and the control of the device.

Among valid alternatives, the photonic microring resonator can be replaced by a chirped Bragg grating, which could be fabricated in a waveguide for photonic integration as well. The grating can be driven with temperature control tuning to change the time delay to be added. A chirped grating has a broad spectral profile (i.e., phase and amplitude versus wavelength) in reflection which shifts with temperature. So, for a fixed wavelength signal, temperature tuning the grating would result in a change in the location within the grating that the signal is reflected from. If, for example, the signal wavelength is set at the middle of the spectral profile then the temperature can be tuned up and down to reduce and increase the delay. The spectral profile would need to be selected so that the signal wavelength remains within the grating profile when tuned.

The reflection profile of a chirped grating can have a relatively constant reflectivity across a good range of wavelengths. Therefore, with a relatively constant reflectivity, at the selected wavelength, when tuning the grating to move that wavelength along the grating profile, it may be possible to get rid of the VOA and have less intensity variation of the delayed signals across the phased-array antenna elements.

8.11 How Optics Can Be Beneficial in Performing a Stable Clock and Frequency Reference

Optically generated RF clocks represent a flexible solution for the most demanding applications. Some interesting research work proposed a functional architecture of beamforming antennas employing mode-locked laser (MLL) as local oscillator and system reference clock [10] and [11], selecting the RF signal after the photodiode detection within several beatings by means of an RF filter, implying a reduced efficiency in the signal generation.

Some research work achieved high technology readiness level (TRL) demonstrating very low phase noise in carrier generation. In the solution proposed in [11] –125 dBc/Hz at 10 kHz and –130 dBc/Hz at 100 kHz, operating at 10GHz were demonstrated. With the same technology in combination with true-time-delay technique, for the phase shifter realization, is possible to reduce the beam pointing error to 0.1 degrees, an order of magnitude lower than the current state of the art precision. Other examples can be found achieving higher precision and higher frequency generation with a TRL to be properly specified [12].

8.12 Conclusions

In future mobile, both LTE and 5G, optics can be beneficial to perform and improve beam precision on single and multiple frequencies. In the last years, the cross fertilization between photonics and radio systems is leading to overcoming the main limitations of the classical radio technologies, setting new paradigms, promising improved performance and new applications, with strong benefits for the fast evolution of new 5G networks.

Photonics-based wireless systems, although still at the research level, are moving toward a new generation of multifunctional systems able to manage the wireless communications with several different frequencies and protocols, even simultaneously, while also realizing communication link analysis operation with high resolution. Photonics matches the new requirements of fast responsiveness of the next generation mobile networks, thanks to its ultra-wide bandwidths and ease of tunability, guaranteeing low footprint and weight leveraging on integrated photonic technologies. Moreover, photonics also allows increased resolution and sensitivity by means of the used inherent low phase noise lasers.

Today's wireless communication systems need specific hardware working at specific radio frequencies, with characteristic signal waveforms and bandwidths, requiring dedicated apparatus for every single application.

Moreover, personal communications will need to exploit new frequency bands as the unlicensed millimeter waveband around 60 GHz. Current electronic devices present increasing limitations and worsening performance as the frequency gets higher. Therefore, in most of the future transceivers for wireless applications, it will be necessary to generate and receive very stable high-frequencies and wideband RF signals by means of reliable transmitters/receivers that also respect constraints in size, weight and power consumption.

Conventional electronic RF transceivers are noisy and suffering from electromagnetic interferences, causing degradation of the sensitivity and of the dynamic range of the whole system. These deficiencies are magnified as wider bandwidth systems are developed.

However, photonics has proved high precision and ultrawide bandwidth [13], allowing the generation of highly stable multiple radio-frequency signals with arbitrary waveforms up to the millimeter waves [14], and their detection and precise direct digitization through demonstrated efficiency improvement of optical to electrical conversions (i.e., without the need to use noisy RF downconversions) [15, 16].

References

[1] Ericsson Mobility Report, November 2018; https://www.ericsson.com/en/mobility-report/reports/november-2018.

[2] Boccardi, F., et al., "Five Disruptive Technology Directions for 5G," *IEEE Commun. Mag.*, Vol. 52, No. 2, February 2014, pp. 74–80.

[3] Zhao, A., and Z. Ren, "Multiple-Input and Multiple-Output Antenna System with Self-Isolated Antenna Element for Fifth-Generation Mobile Terminals," *Microwave and Optical Technology Letters,* 2018.

[4] Ghelfi, P., et al., "Photonic Generation and Independent Steering of Multiple RF Signals for Software-Defined Radars," invited paper, *Optics Express,* Vol. 21, No. 19, September 2013, pp. 22905–22910.

[5] Yi, X., T. X. H. Huang, and R. A. Minasian, "Photonic Beamforming Based on Programmable Phase Shifters with Amplitude and Phase Control," *IEEE Phot. Technol. Lett.*, Vol. 23, No. 18, September 2011.

[6] Hunter, D. B., M. E. Parker, and J. L. Dexter "Demonstration of a Continuously Variable True-Time Delay Beamformer Using a Multichannel Chirped Fiber Grating," *IEEE Trans. Microwave Theory Techn.,* Vol. 54, No. 2, February 2006.

[7] Kaman, V., et al., "A 32-Element 8-Bit Photonic True-Time-Delay 288 3-D MEMS System Based on a 288 Optical Switch," *IEEE Phot. Technol. Lett.,* Vol. 15, no. 6, June 200.3

[8] Piqueras, M. A. G., et al., "Optically Beamformed Beam-Switched Adaptive Antennas for Fixed and Mobile Broad-Band Wireless Access Networks," *IEEE Transactions on Microwave Theory and Techniques,* Vol. 54, 2006, pp. 887–899.

[9] Zhuang, L., et al., "Single-Chip Ring Resonator-Based 1x8 Optical Beamforming Network in CMOS-Compatible Waveguide Technology *IEEE Phot. Technol. Lett.,* Vol. 19, August 2007, pp. 1130–1132.

[10] Ghelfi, P., et al., "Photonic Generation and Independent Steering of Multiple RF Signals for Software Defined Radars," *Opt. Exp.,* Vol. 21, No. 19, 2013.

[11] Adams, D. B., and C. K. Madsen, "A Novel Broadband Photonic RF Phase Shifter," *Journal of Lightwave Technology,* Vol. 26, No. 15, 2008.

[12] Tsokos, C., et al., "Analysis of a Multibeam Optical Beamforming Network Based on Blass Matrix Architecture," *Journal of Lightwave Technology,* Vol. 36, 2018, pp. 3354–3372.

[13] Capmany, J., and D. Novak, "Microwave Photonics Combines Two Worlds," *Nature Photonics,* Vol.1, 2007, pp. 319–330.

[14] Ghelfi, P., et al., "Phase Coding of RF Pulses in Photonics-Aided Frequency-Agile Coherent Radar Systems," *IEEE J. Quantum Electronics,* Vol 48, No 9, 2012, pp. 1151–1157.

[15] Laghezza, F., et al., "Jitter-Limited Photonic Analog-to-Digital Converter with 7 Effective Bits for Wideband Radar Applications," in *2013 IEEE Radar Conference (RadarCon13),* 2013, pp. 1–5.

[16] Khilo, A., et al., "Photonic ADC: Overcoming the Bottleneck of Electronic Jitter," *Opt. Express,* Vol. 20, No. 4, 2012, pp. 4454–4469.

CHAPTER 9

Photonic Applications for Radio Systems and Networks

9.1 Introduction

Between 2018 and 2024, total mobile data traffic will increase up to five times, and 25% of mobile traffic will be transported by 5G networks by the end of the period [1]. With 5G networks, there is an opportunity for telecom and network operators to push for new business models, and so will be the next evolutionary steps. Leveraging on the new business trends, Telecom and Datacom networks evolve with the creation of dedicated new ecosystems bringing different partners together, each one offering different technological services, combined for the first time in an innovative process of ICT growth.

5G networks are aimed to introduce 100-times energy efficiency better than 4G networks, with significant impacts in the CO_2 emission footprint, in favor of sustainability and green operability [2]. As the energy dissipation for handling information has risen to environmentally significant levels, continuing to handle the exponential growth in information traffic cannot be possible without the use of new technologies that significantly reduce the energy per processed and transmitted bit. An important role in 5G evolution can be ascribed to the cloud infrastructure supported by evolved IP delivery platforms bringing an increasing amount of new Telco and Datacom grade services.

One of the major impacts for network service providers leveraging on new cloud infrastructures is the backhaul of traffic upgrade, as it is rapidly growing due to the introduction of high-speed data services. Therefore, there is the need to add more transmission capacity to the backhaul of wireless communications networks, to handle in parallel the increasing data traffic and other traffic such as voice. A solution that has been considered is to use both microwave and fiber-optic backhaul in the lower radio access network of a wireless communications network.

However, while the fiber-optic backhaul provides a much higher traffic transmission capacity than the microwave backhaul, the costs of implementing a fiber-optic backhaul are considerably higher than a microwave backhaul; the major cost impact is in digging. A figure of merit that could be a determining factor in whether one technology is preferred over the other is the cost per bit transmitted of the whole transmission system: Transceiver plus propagation media (e.g., air, fiber) and logistics costs, included communication license, system infrastructure (already existing, or to be updated or build from scratch), power-supply system costs, and others.

The chapter is organized as such: Section 9.2 introduces the IP core network in 5G, focusing on high-speed line interfaces, and mentioning the main requirements of low power consumption; Section 9.3 is focused on the use of photonics in the transceiver, highlighting the benefits compared to electronic solutions; Section 9.4 discusses the use of integrated technologies, with a specific focus on wavelength division multiplexing systems; and Section 9.5 gives a special mention on the importance of pluggable modules.

9.2 The IP Core Network Evolution

In 5G systems, and following development, a significant transformation takes place in the central part of a telecommunications network that provides various services to final users connected by the access network; this is the IP core network. The IP core has the main purpose of providing network services to connect the primary nodes in the edge network with the access network on the one hand, and with the rest of the network (i.e., core network) on the other, as shown in Figure 9.1.

Following the above outline for the IP core platform evolutionary story, two main points can be highlighted where optics can introduce relevant innovation. The progress of line cards toward a huge bit rate, to perform packet processing and switching new generation of photonic integrated circuits, can be introduced to scale up performance and reduce costs, power consumption, and footprint. The capacity increment for interconnections between huge capacity line cards and between cards and backplane: new performing applications can leverage on the last technological advances of photonic integrated transmission interfaces.

Huge traffic throughput is expected in the IP core network with 5G and beyond technologies. An evolved concept of the core router platform applied in this scenario can be the one equipped with line cards, optical backplane, and interconnections performing high-speed transmission, routing, and interconnection. As reported in recent relevant works, the core router can be efficiently operated in tandem with an optical switch-based solution to support broadband interconnections [3].

Looking at the technology advances in datacenters, big optical pipes are enabled inside the network to perform traffic offloading for example during data storage mirroring. In some examples, an optical switch can be added in the system to perform the packet offload and avoid power hungry processing in the interested routers. If silicon photonics is used to realize the high-speed optical router interfaces and the optical switch structure, the cost and the power consumption of the entire system are significantly reduced while still supporting heavy operations in intradatacenter communications [4].

In the case of IP core network for telecom applications, the definition of the packet size is crucial for the definition of a core router architecture. Every core router section is dimensioned on the analysis of real traffic of the network segment of interest. The core router design must be optimized for IP traffic routing lowering the probability of network congestion.

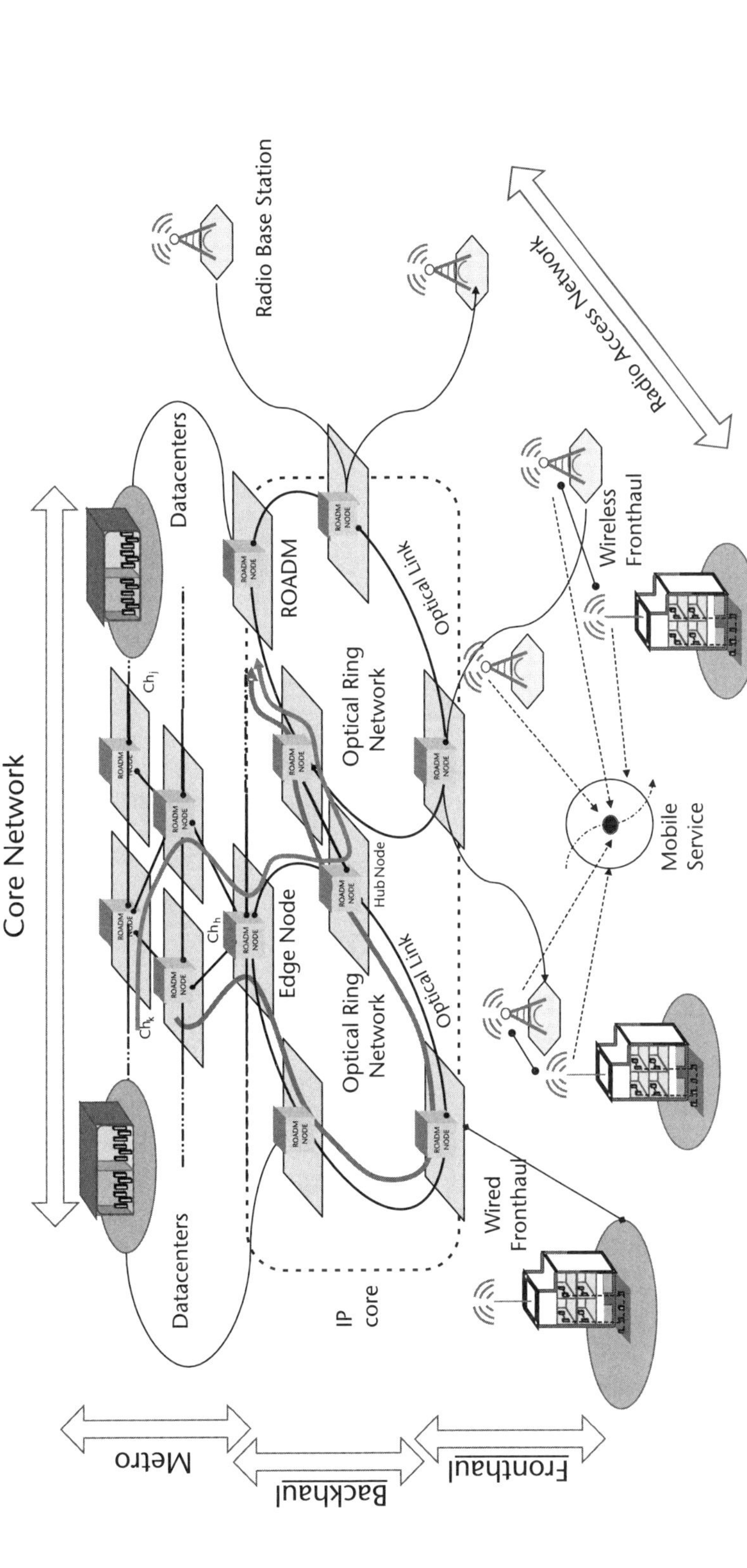

Figure 9.1 Access and core network with the IP core network in the central part of the network.

There exist two possibilities to evaluate the core router performance and define the NPU design. The router can be dimensioned as

1. All packets have the minimum acceptable size (i.e., ACK packets for IP network). The router parameters are set in the worst-case scenario;
2. Packet size slightly varies from the mean packet frame in the network of interest.

The first approach leads toward an overdimensioned router architecture with respect to real traffic requirements to reduce to zero the probability of traffic congestion. The main limitations of this approach are only due to the strategy executed by the line card in handling packets harvested and distributed by the line interfaces. The second approach leads to an optimized core router system, equipped with the real required resources for routing packets with an acceptable probability of network congestion. To take no risk of traffic loss, the line card set the routing parameters basing on real traffic analysis practically establishing an acceptable level of congestion. Substantially the second approach optimizes the cost and the power consumption of the entire IP core router.

9.2.1 High-Speed Line Interface for IP Core Router Line Cards

It is worth noting that to obtain a consistent reduction of power consumption while enhancing router line card performance by choosing the proper technology, further format enhancement of the line interfaces, and the backplane interfaces must be adopted for high-speed connections. The data traffic in IP core or datacenters can be handled by electrical interconnects. However, they have intrinsic high channel losses of strip-lines and wires (e.g., based on gold or copper), pin count constraints, and require sharp impedance matching in the used hardware to avoid energy leakage and unwanted bandwidth limits.

The alternative based on the use of optical interconnects exhibits low loss at high symbol rates, long bridge length, and immunity to electromagnetic interference. Optical interconnects do not require impedance matching to be effective in terms of energy efficiency and interconnection bandwidth. Substantially, optical interconnections can reduce the cost, the power, and the complexity compared to the electronic link mainly in case of interdatacenter, rack-to-rack, board-to-board, and module-to-module interconnections.

A significant contribution to energy consumption can be ascribed to the clock and data recovery operations, and time multiplexing in interconnect links, line coders, and serialization and deserialization (SERDES) circuits (for handling clocking, data retiming, and time multiplexing issues). Optical interconnect technology has been identified as an alternative solution to overcome the foreseeable scaling limitations of conventional interconnects.

To outline the energy targets for the technological advances in system interconnections in IP core network and datacenter, Table 9.1 reports the main value references for different types of the interconnection link.

Depending on the link length, different energy consumptions per transmitted bit are required to support high-speed interconnection for different applications:

Table 9.1 Energy Targets for Interconnection [3, 4]

Interconnection	*Distance*	*Energy per Bit*
Interdatacenters/IP core nodes	1–100 km	<10 p J/b
Rack-to-rack	1 m–2 km	<1 pJ/b
Board-to-board	0.3–1 m	<1 pJ/b
Module-to-module	5–30 cm	< 0.5 pJ/b
Chip-to-chip	1–5 cm	<0.1 pJ/b
Multicore	<1 cm	<0.01 pJ/b

- Interdatacenter or IP core nodes interconnection, where huge traffic flows determine high bit rates transmission and switching;
- Rack-to-rack interconnection, where point-to-point interconnections are required at a high bit rate;
- Board-to-board interconnection, where energy-efficient transmissions are needed allowing for high data throughput;
- Module-to-module interconnection, where fully integrable onboard solutions are needed to optimize energy and footprint of the router cards;
- Chip-to-chip interconnections, typically realized with electrical wires or strip-lines until a certain interconnection capacity per distance product is achieved in the card design;
- Multicore interconnections, where very short reach interconnections among different ASICs on a router card are typically realized with electrical wires or strip-lines.

9.3 The Use of Optical Modules in Router Cards

With reference to photonic transceivers, the wavelength band of operation for Datacom and Telecom applications can be in the O-band (i.e., around 1,310 nm) [5] or in the C-band (i.e., around 1,550 nm) using the short and medium optical link (from meters to tens of kilometers) [6].

The solutions rely on two different approaches: Pluggable and integrated onboard transceivers. The solution based on the use of pluggable optical modules can be valuable if these can scale capacity while keeping constant the hardware footprint and power consumption depending on the use case. The pluggable solution must be designed to have small form factor, to improve the port density on the frontend of a router board (e.g., a WDM transponder or a transceiver card). To this aim, new basic technologies are required enabling more functional transceiver design and enhancing the bandwidth per link-length product of an optical connection.

On the other side of integrated onboard solutions, full integration can lead to better performance in terms of energy efficiency and consequently cost. For instance, the integration of an optical source like a laser over the same silicon substrate hosting a microprocessor or a digital signal processor module, stabilized in temperature and protected by environmental vibrations, energy optimization of the overall connection system can be reached. The possible circuit design simplicity and energy efficiency are fundamental to dramatically reduce the overall cost per bit.

9.3.1 Expected Evolution in IP Core Network with Integrated Onboard Modules

A possible technological advance oriented for IP core networks and datacenter applications is represented by the novel approach of integration on-die in a 3-D configuration to realize onboard optical modules. The 3-D integration solution enables the interconnection shortening between different parts of a router card consisting of ASICs, memory banks, and cross point switches. Three-dimensional face-to-face integration allows for coexistence on the same platform of photonic integrated circuits and electronic integrated circuits; the first to modulate and process light for the chip-to-chip or multicore interconnections, the latter to process and store data and increase the computational capacity. The solution leverages on the exploitation of the silicon interposer layer to hold electrical or optical interconnections at a high bit rate between onboard application specific integrated circuits (ASICs) depending on the application and link-length. The electrical interconnections can be realized through a dense array of micro copper pillars (i.e., low parasitic capacity and inductance, below tens of femtofarads and picohenries, respectively) allowing for a direct connection at high bit rate (e.g., 25 Gb/s) of an ASIC with the host board (the printed circuit board, or PCB) or with the photonic integrated circuits fixed in the optical interposer layer for 3-D electrical to optical conversion.

The interconnection solution throughout micro copper pillars requires that PIC and EIC are integrated last after the other on the common PCB, optical waveguides can be implanted in the optical interposer layer for the chip-to-chip interconnection or multicore interconnection. An example of 3-D integration is reported in Figure 9.2 with electrical interconnections and with optical interconnections hosted in the silicon interposer.

Leveraging on the proposed 3-D ASIC integration significant power saving can be achieved by

- Reducing the number of interconnections between ASIC's and switches with the exploitation of well-performing (i.e., optimized coupling) electrical or optical connections;
- Lowering the electrical line length for the chip-to-chip interconnection and energy feeding, thus enabling the elimination of power consuming electrical drivers;
- Exploiting optical line interconnections are considered in the silicon interposer the energy saving is at least 30% over current since the photonics wiring will reduce the power consumption generated by electrical wiring [7].

It is worth noting that the last decades the progress of conventional CMOS electronics has been governed by Moore's Law, stating that the number of transistors on a microprocessor chip will roughly double every two years. This means that

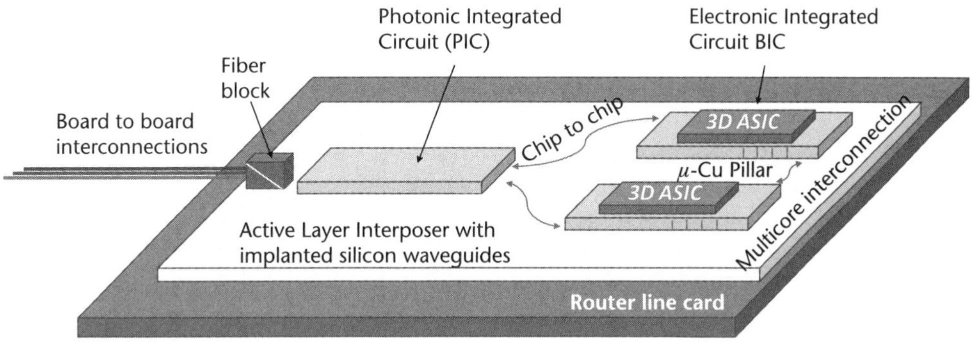

Figure 9.2 Example of 3-D integration with optical interposer and EIC and PIC [14].

more and more transistors and integrated circuits can be placed onto smaller areas. However, in the last years, the doubling has started to slow down due to heat dissipation and technology scaling challenges from a large number of transistors and circuits in the small chip area [8]. As the device dimensions, such as the gate oxide thickness, have been reduced to several atomic layers, tunneling and leakage current become significant. The limit is expected to be reached for gate lengths around 5 nm because of increasing leakage currents [9], known as gate leakage limitation [10]. Some solutions to this problem have already been proposed, such as 2-D and 3-D gates [11] and FinFET gates, as already adopted by major technology vendors [12].

For instance, the technology 3-D gate can be used to increase the ASIC performance of 37% and lowering the power consumption of 50% [13]. This technology is available and is expected to be used and optimized for mobile devices applications and eventually for IP core networks and transport networks by implementing the 3-D face-to-face integration between photonic and electronics. Thus, significantly improving energy efficiency and cost optimization in the overall network while scaling up with interconnection and computational capacity.

9.3.2 High-Speed IC Interconnection

In the aim to evolve an IP core router toward advanced solutions, an innovative approach can be suggested, nowadays consolidating as 3-D integration of vertically stacked integrated circuits connected using through silicon vias (TSV) technology for high-bandwidth connections (e.g., 25 Gb/s). This is the most promising technique for low power consumption and reduced costs and footprint and is depicted in simple implementation schemes shown in Figure 9.3.

The same technique can be used to realize ASIC and FPGA as well, which can be dedicated to implement the part of the router system with programmable

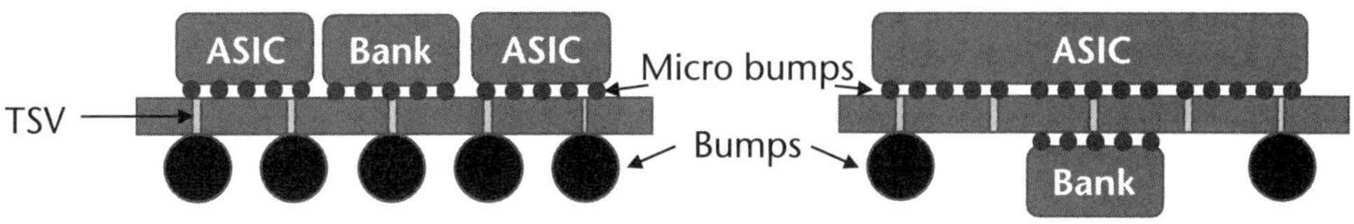

Figure 9.3 Examples of 2.5/3-D integration to realize the intrachip interconnection.

functionalities introduces a silicon interposer to optimize connections for intra-chip communications with optics. The 3-D ASIC integration can enhance the throughputs of interconnects coping with the further demand increase in pin density and data transfer rate. This solution can be realized with electrical connections, but introduces in the near-term applications some important limitations imposed on speed, density, and power due to the limits of electronics. Therefore, for near-term solutions with interconnects, on-chip ASIC can be achieved in an optimal manner. With this optical approach interconnection, optical waveguides can be designed in the silicon photonic interposer and optical engines can be efficiently integrated on the silicon platform if realized in silicon photonics. An example is reported in Figure 9.4.

In the silicon photonic interposer are implanted not only the optical waveguides but also the optical modulators (e.g., MZMs), the photodetector for optimized electro-optical and optoelectronic conversion, respectively, at high bit rate (e.g., 25 Gb/s), electrically driven throughout TSVs.

An interesting scheme example for a 3-D integrated circuit realization including the silicon interposer encompasses distributed optical light sources on the line card, mounted in a separate section not interested by 3-D integration process where is possible to optimize the temperature stabilization of the used laser array. [15]. The light sources are conveniently realized in silicon photonics using III-V materials (e.g., indium-phosphide, indium-gallium-arsenide).

For instance, 4x25Gbps transceivers [16] realized in CMOS photonics can be designed for backplane interconnection and can be used on the next 400 GB line card. Further improvement in terms of performance and power consumption can be achieved by proposing alternative solutions based on high-speed FSO interconnections performed with submillimeter optical terminals realized in silicon photonics [17].

Considering all the above key silicon photonics technologies the target of obtaining power consumption below 1 W/Gb is possible and competitive solutions can be designed [3, 4].

9.4 Integrated Photonics Technology for Multiwavelength Line Cards

Ultrahigh line rates (e.g., 100, 400 Gb/s, 1Tb/s) in next-gen IP core infrastructures can be expected to be supported by connections composed of multiple carriers in optics generated by a single line card in a router. The increase of traffic demands

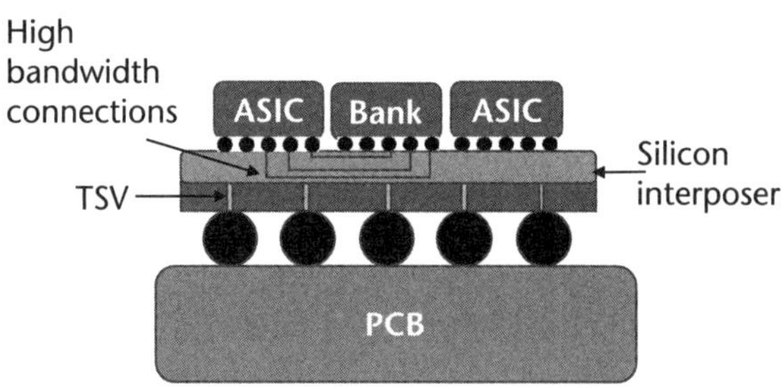

Figure 9.4 Example of high-bandwidth connections implanted in the silicon interposer.

and dynamicity have directed telecom system vendors research and development towards novel classes of transceiver or transponder cards in optical networks. The main requirement for operators is to have line cards providing several functionalities, as the support of multiple line rates (e.g., from 100Gb/s to 1Tb/s) and link distance adaptation with acceptable system margins [18–20].

The aforementioned line cards are still at research level and are designed to be able to generate multiple optical flows (referring to subcarrier as a flow unit), that can be flexibly directed toward different paths and destinations, depending on traffic requirements. Considering reach adaptation, this can be achieved by trading off spectral efficiency and transmission robustness. Two approaches to reach adaptation have been considered in the literature:

1. Modulation format adaptation [18–21];
2. Code adaptation [22–23].

According to the former, by fixing the line rate, the more spectral efficient polarization multiplexing 16-quadrature phase shift keying (PM-16QAM) is less robust than the less efficient polarization multiplexing quadrature phase shift keying (PM-QPSK), which can reach larger distances. According to the latter, more redundancy (coding) is inserted for paths characterized by a lower optical to signal noise ratio. An example of code-adaptive transmission technique is time-frequency packing (TFP) [22–23].

Each optical channel can be transmitted on a different carrier which is typically generated by a dedicated laser source [18], alternatively, all carriers can be obtained through a single multiwavelength source on the line card [19] (i.e., a source able to generate multiple carriers starting from a single laser source). The second solution requires further research and engineering development but allows to effectively reduce the number of lasers onboard and in the network.

The proper choice of multiple carrier generation may depend on the wavelength instability of the used commercial laser sources (e.g., 1 or 2 GHz of frequency instability can be tolerated in WDM system realization on the ITU-T G.694 grid). Carrier frequency overlapping may occur, causing strong signal degradation especially if spectral-efficient transmissions (e.g., Nyquist [24–28] and Faster-than-Nyquist transmissions [29–33]) are performed in the system to significantly scale up with transmission capacity.

The multiwavelength laser source can be the best choice if the technological solution involved does not imply a carrier overlapping, generating a wavelength comb with a well-fixed carrier spacing. Researchers and industry are investigating commercial solutions for multiwavelength source considering it as a practical solution to generate multicarrier transmission reliably to scale up with transmission capacity on a single fiber.

The main drawback of the multiwavelength source use on a line card is the need to select each spectral component to be modulated by the specific optical modulator. The more spectral efficient is the transmission technique used, the smaller the wavelength comb spacing, the sharper should be the optical filter to select the spectral component to be modulated.

An example of a line card supporting multiple line rates and link distance adaptation with multiple carrier transmission has been proposed in [18] where silicon photonic integration is used to realize transceivers and electro-optical interfacing. In a dedicated space of the line card, not interested by the silicon photonic electro-optical integration onboard, a multiwavelength source is used to generate the optical carriers. General schematic of the referenced line card is reported in Figure 9.5.

The number of carriers depends on the used electronics setting the transmission baud rate and the overall capacity to be transmitted in the line.

The referenced line card architecture has a modular structure for a flexible and multifunctional platform with the ability of adaptation to IP Core network scenario [18]. In the referenced model clients from electronic layer (e.g., optical transport network, OTN; Ethernet) enter the line card to modulate optical several carriers for line transmission. The carrier generation module can generate up to 10 wavelengths being transmitter and receiver local sources. The carrier modulation at the transmitter side is realized by gathering the client traffic (i.e., Ethernet, OTN traffic) and processing the relevant data through a dedicated DSP for encoding, pulse shaping, and predistortion. The digital data are spectrally shaped and converted to analogic samples by a dedicated DAC and then enter the optical modulators. At the receiver side, each transceiver uses coherent detection, ADC, and then digital processing for optical impairment mitigation, clock recovery, synchronization and others. Polarization multiplexing is used to double the bit rate, then optical wavelengths can be coupled or multiplexed and launched in the network through add/drop ports of an optical node (see Chapter 6). The optical transceivers can be integrated on board through silicon photonic integration or accommodated in pluggable modules which can be added, removed or replaced depending on network needs. Each pluggable module can be designed to generate up to 400 Gb/s, to reach high line card throughput, that may exceed 1.6 Tb/s data traffic.

9.4.1 Photonic Integrated Transceiver

New, effective solutions for integrating the optical components in a photonic integrated circuit (PIC) are on the stage that aim to save significant cost, power, and footprint [34]. New photonic integrated transceivers can be operative in different telecommunication bandwidths [34] and different network segments (e.g., metro, metro regional, core) [35]. The development of CMOS photonics, in conjunction with advanced materials, (for example, silicon-organic hybrid (SOH) structures [35]) can lead to the optimization of performances and energy efficiency.

The significant driving voltage reduction (from 4–5V to lower than 1V) and the use of existing CMOS production lines enable high yield and cheaper photonic solutions [35]. An effective approach can be the production of photonic integrated

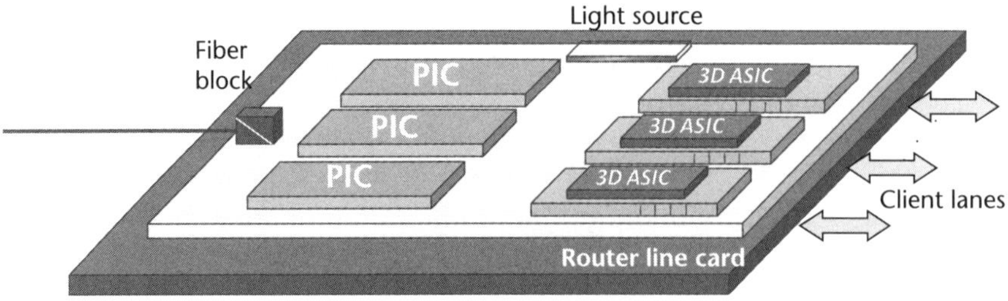

Figure 9.5 Multiwavelength card to significantly scale up with capacity in IP core network [18].

transceivers throughout the 45 nm CMOS production process over a 300-mm wafer scale. This can introduce significant cost savings from leveraging on high yield well-established production lines. The photonic transceivers can be designed to perform several types of high-level modulation format effective in transport links (e.g., QPSK, M-QAM) and data links (e.g., PAM-4).

The mentioned model for multicarrier transmission in IP core network strongly relies on three main technological advances corresponding to three different functional blocks: The optical transmitter, encompassing a modulator and coherent detector, which can be based on silicon photonics [34]; application-specific integrated circuitry (ASIC) for signal processing in transmission and detection, which is based on the latest developments in CMOS electronics [13]; and multicarrier light source module realized with dedicated fabrication processes, such as those based on the use of III-V materials [36]. Each functional block can be designed and produced with dedicated value chains to optimize performance, power, and cost.

For electronics, the most consolidated technological advance is represented by the CMOS process evolution below 10 nm electronics, which aims to leverage very large wafer scale for production; over 400 mm wafer scale for the realization of an ASIC with several hundreds of millions of gates for high-speed signal processing, operative at 50 Gbaud/s (symbol rate). In such a way, the ASIC can encompass DSP-enabled adaptive capabilities and high-speed DAC and ADC (112-128 Gsample/s and 12 digits resolution).

Scaling below 10 nm CMOS with the gate dimensions and using FinFET [12] technology to achieve baud rate up to 56–64 Gbaud/s is feasible. Thanks to the baud rate upgrade, the next generation terabit transponders can offer higher capacity, higher spectral efficiency and longer distances by performing multiple modulation formats, multiple baud rates, advanced coding, and flexible optical resource allocation.

The aimed advantage with the symbol rate increment is to reduce the number of needed carriers in the multiwavelength source, simplifying the overall line card architecture. Power consumption optimization can be also performed; by increasing the baud rate and for a given modulation format (e.g., QPSK), the number of onboard of transceiver modules can be reduced to achieve more than 1.6 Tbit/s line card data throughput.

Still, technological challenges should be faced for sustainable high-density integration and packaging on a line card. For instance, physical implementation limits and most of the raised complexity can be mainly ascribed to the bandwidth narrowing effect of cascaded DAC, driver, and modulator at the transmitter and of the cascaded detector, TIA, and ADC at the receiver side. Furthermore, the required techniques to recover from linear and nonlinear distortions and propagation impairments become even more challenging [37].

9.4.2 Multicarrier Light Source

The multicarrier light source can be an array of N tunable lasers, connected to the high-speed transceivers via a possible silicon interposer, or PCB, or backplane. Alternatives to the use of discrete N independent DFB lasers are possible: Single die monolithically integrated lasers or multiwavelength locked lasers [38, 39].

Among all possible realizations, the full integration of light sources onboard in a cost-effective and low footprint can be achieved throughout III-V material, leveraging on the best yield so far achieved in III-V epitaxial layers grown on a 150 mm indium phosphide wafer [40] adopted for the realization of laser array for subcarrier generation. A common packaging, temperature stabilizer, and wavelength control enable cost-effectiveness for the comb generation with acceptable performance in terms of phase noise and wavelength stability.

As an alternative approach, the use of a scheme which generates multiple optical carriers starting from a single laser has been studied and several examples of multiwavelength carrier generation have been reported [41].

The multiwavelength source solution has advantages and drawbacks. On one hand, costs and power consumption can be optimized and carriers can be phase-locked so that mutual crosstalk can be counteracted with reduced complexity signal processing. On the other hand, depending on the chosen implementation, carrier wavelengths are relatively fixed, thus reducing the transponder flexibility [38].

A crucial feature that can be easily enabled through new technological approaches is the programmability of the laser comb: each individual subcarrier is required to be properly selected for independent modulation. To add wavelength flexibility, the comb must be tuneable to properly set carrier spacing depending on the wanted spectral efficiency or the quality of the optical filters used to select each wavelength in the comb, and not least on the number of carriers [38]. To be operative with those features, the multiwavelength source is equipped with tuneable and selectable filters [38]. Possible solutions can result in a not-so-cost-effective and power efficient application with technical limitations such as filter bandwidth and shape [42–45]. An alternative can be to limit the number of wavelengths, as proposed in [41].

9.5 Optical Interconnection with Pluggable Modules

Recent research work has been focused on defining the optimal, critical length for certain line rates for which optical interconnects can indeed offer a competitive substitute over the corresponding electronic-based ones. It has been stated that the critical length beyond which the optical interconnects become advantageous over electrical interconnects is approximately one-tenth of the chip edge length at the 22 nm technology node (using a chip edge length of 17.6 mm) [3, 4, 7, 8].

Another similar study shows that on-chip electrical interconnects suffer from losses at lengths beyond 200 cm, for the 22 nm node technology and hence it would be better substituted by optical interconnects [3, 4, 7, 8].

For board-to-board systems, when data rates become higher than 20 Gbps, optical interconnects should replace electrical ones, since the latter would need power-consuming amplification to be used to compensate for transmission loss [3, 4, 7–9].

The alternative technologies in optics can rely on coherent or direct detection transmission techniques. Both solutions are not scalable with a signal bit rate increase in a cost-effective way.

The first solution requires complex and power demanding transmitter and receiver realizations with the necessary use of DSP circuitry to recover information from QPSK or M-QAM formats, or polarization multiplexed signals increasing signal capacity but detectable with the heterodyne structure of a complex optical receiver [46]. The solution has high realization costs, and today can be used only over long bridged distances (e.g., more than a hundred kilometers), only where the cost is not the more stringent parameter.

The second approach, based on direct detection, practically scales up the capacity of the transmitted signal by increasing the number of amplitude levels (M) by PAM – M-PAM – formats and uses a photodiode in combination with a simple but power demanding DSP to recover information functional for short distances (e.g., up to few kilometers).

Both solutions are designed to be not tied with the cost increase of electronics adaptation to optics versus capacity scaling of a transmitted signal, but the implementation costs are still high to be applied in RAN. Current electronics can be scaled to operate up to 60 GHz with reliable realizations, but the integration cost with optics can be prohibitive for radio access applications. A not negligible cost can be ascribed to the need for having a functional packaging of the optical interface with standardized footprint and power consumption. For instance, encapsulating 100 Gbit/s complete transceiver in a small standard format like CFP8 (i.e., 21-mm dimension and 6W overall allowed power consumption in MSA standard) is really challenging if the DSP circuitry is included. A viable solution can be to perform 100 Gbit/s with the NRZ OOK basic format, avoiding codification on a multilevel format through a DSP and by applying direct detection.

If this optical transmission is performed in the C telecommunication bandwidth, there are physical limitations to be counteracted. The limitation can be ascribed to the detrimental propagation effects due to CD and PMD of the fiber link on the transmitted signal. This effect should be compensated to recover information by scaling up the signal bandwidth. Errors are caused mainly by the consequent aliasing. Available compensation techniques are not cost-effective and can be performed in the optical domain or by a post-detection DSP, causing high propagation losses to be recovered in the first case, or requiring a complex and high performing electronics in the latter case.

9.5.1 Pluggable Module Form Factor

Having a heterogeneous structure to realize the optical elements, like modulators, photodetectors, filters and switches, and optical sources cause a cost increment of the integrated transceivers due to adaptation procedures needed to align an external optical source to the available photonic circuitry.

Because of having the light sources heterogeneous (e.g., based on III-V materials) or external to the silicon-based photonic circuitry, the final cost of a broadband transceiver or switch depends mainly on the packaging realization issues in the final product. Packaging focuses on the design, assembly and component protection. The objective is always to balance between flexibility, performance, reliability and low costs, and to increase the density and the number of system functionalities. Reducing this percentage is fundamental to reduce the overall cost of the final component.

The footprint reduction of pluggable modules can help to increase optical interface density on the router card frontend, and consequently to optimize the overall system cost.

A multisource agreement, called "C form-factor pluggable" (CFP), was established to produce common form-factor pluggable transceivers. Primarily, the agreement was defined for 100 Gb Ethernet systems, then updated up to 400 Gb Ethernet systems. From CFP to CFP4 form factor, the space occupancy is reduced by a factor of four. Other relevant formats smaller than CFP4, is QSFP28 multisource agreement which offers a higher port density [47].

In terms of bandwidth density, the form factor CFP2, specified as form-factor of 1/2 in size of the CFP module, allows for dealing with up to four 100 Gb/s transceiver modules plugged on a single router card to process up to 400 Gb/s traffic rate. The same can be analyzed for CFP4 solution by almost doubling the overall traffic rate on the router card (i.e., 800 Gb/s). Up to 8 CFP4 line interfaces can be hosted in the router card to achieve the needed traffic throughput.

To classify the pluggable form factors, the figure of merit for already existing and underdevelopment modules can be the relative bandwidth density as the ratio between data rate, in Gigabit, and module width, in millimeters. In Figure 9.6 the relative bandwidth density for the different pluggable module form factors is outlined [47].

CFP modules are capable to support 10-channel modulated at 10 Gb/s (10x10G) as well as a 4-channel setup divided into 25G signals (4x25G). QSFP only supports a 4-channel setup, in QSFP28 module each lane can carry up to 28G Ethernet (4x28G). Hence, CFP modules are more versatile than QSFP28 transceivers in terms of channel setups.

Pluggable modules (e.g., CFP, QSFP multisource agreement) available on the market can be used in radio application, to interconnect the BBU to the RRU in

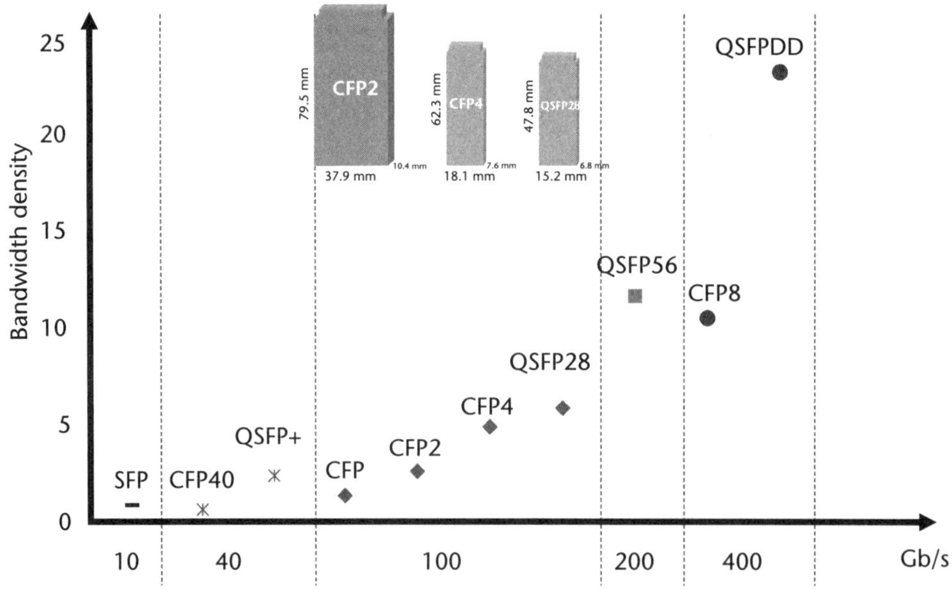

Figure 9.6 Relative bandwidth density of various optical modules [47].

fronthaul networks, or in IP core networks on the line cards processing up to 800 Gb/s overall traffic rates.

9.6 Conclusions

Current and new applications for the IP delivery platforms, like those targeted for IP Core network, require a crucial upgrade of the entire core system.

As far as the traffic throughput increases in the network segments where an IP router is operative, there exists the incumbent need to increase core system performance and functionalities. Great help can be represented by the last technological advances in optics.

Optical research activities can be carried on pursuing an effective solution in updating new generation core router platforms. The crucial point of interest is to enhance the traffic throughput sustainable by the core router. The solution can be mainly oriented to the realization of high-speed line card with efficiently scaled cost, power consumption and footprint throughout the exploitation of new technological advances.

New solutions are on the stage today that can influence the design of new core routers in the 5G scenario and its developments, and are mainly based on silicon photonics technologies. For instance, to perform high-speed packet processing (1.6 Terabit full duplex and beyond), 3-D optical ASIC with closer optical photonic integrated circuits can be designed to enhance the matching interconnectivity between electrical and optical units.

To perform high-speed short reach interconnection between cards or different routers, optical transmission can be implemented with silicon photonic integrated circuits, to overcome the limits of the use of electrical lines to handle huge bite rate (400 Gb/s and beyond) bridges, and hundred centimeters or meters length. For longer bridges (e.g., rack-to-rack or interoptical node interconnection) integrated onboard photonic modules or pluggable modules at high bitrate can be used with upgraded transmission capacity at 400 Gb/s. Dedicated line cards hosting more than one optical module performing spectral efficient transmission (e.g., Nyquist or faster-than-Nyquist coherent transmission) can be realized to further improve system performance keeping sustainable energy consumption.

References

[1] Ericsson Mobility November 2018, https://www.ericsson.com/en/mobility-report/reports/november-2018

[2] file:///C:/Users/eantder/Desktop/Graphene/Spear%20Head/ericsson-sustainability-and-corporate-responsibility-report-2017.pdf, Ericsson Sustainability and Corporate Responsibility Report 2017, May 11, 2018.

[3] Kachris, C., and I. Tomkos, "A Survey on Optical Interconnects for Datacenters," IEEE Commun. Surveys and Tut., 2011.

[4] Yoo,S. J. B., "The Role of Photonics in Future Computing and Datacenters," IEICE Trans. Commun., Vol. E97-B, No. 7, July 2014, pp. 1272–1280.

[5] Denoyer, G., et al., "Hybrid Silicon Photonic Circuits and Transceiver for 56Gb/s NRZ 2.2km Transmission Over Single Mode Fiber," 2014 The European Conference on Optical Communication (ECOC), Cannes, 2014, pp. 1–3, doi: 10.1109/ECOC.2014.6964262.

[6] Marxer, C., et al. "Vertical Mirrors Fabricated by Reactive Ion Etching for Fiber Optical Switching Applications," in Proceedings IEEE the Tenth Annual International Workshop on Micro Electro Mechanical Systems. An Investigation of Micro Structures, Sensors, Actuators, Machines and Robots, IEEE, 1997, pp. 49–54.

[7] Arakawa, Y., et al., "Silicon Photonics for Next Generation System Integration Platform," IEEE Communications Magazine, March 2013, pp 72–77.

[8] Waldrop, M. M., "The Chips Are Down for Moore's Law," Nature, Vol. 530, No. 7589, February 2016, pp. 144–147.

[9] Iwai, H., "Roadmap for 22 nm and Beyond," invited paper, Microelect. Eng., Vol. 86, Nos. 7–9, 2009, pp. 1520–1528.

[10] Beausoleil, R. G., et al., "Nanoelectronic and Nanophotonic Interconnect," Proc. IEEE, Vol. 96, No. 2, February 2008, pp. 230–247.

[11] Cartwright, J., "Intel Enters the Third Dimension," Nature, May 2011.

[12] Nitta, C., M. Farrens, and V. Akella, "DCOF—An Arbitration Free Directly Connected Optical Fabric," IEEE J. Emerg. Sel. Topics Circuits Syst., Vol. 2, No. 2, June 2012, pp. 169–182.

[13] Morris, J., "Why Intel's 22nm Technology Really Matters," http://www.zdnet.com/blog/computers/why-intels-22nm-technology-really-matters/5703.

[14] Boeuf, F., et al., "Recent Progress in Silicon Photonics R&D and Manufacturing on 300mm Wafer Platform," 2015 Optical Fiber Communications Conference and Exhibition (OFC), Los Angeles, CA, pp. 1–3, 2015.

[15] Arakawa, Y., et al., "Silicon Photonics for Next Generation System Integration Platform," IEEE Communications Magazine, March 2013, pp 72–77.

[16] Guckenberger, D., et al., "Advantages of CMOS Photonics for Future Transceiver Applications," 36th European Conference and Exhibition on Optical Communication, ECOC 2010, Torino, Italy, paper TU.4.C.2.

[17] Rachmani, R., and S. Arnon, "Server Backplane with Optical Wavelength Diversity Links," J. Lightwave Technol. Vol. 30, 2012, pp. 1359–1365.

[18] Cugini, F., et al., "Demonstration of Flexible Optical Network Based on Path Computation Element," J. Lightwave Technol., Vol. 3, No. 5, 2012.

[19] Jinno, M., et al., "Multiflow Optical Transponder For Efficient Multilayer Optical Networking," Communications Magazine, 2012.

[20] Cai, X., et al., "Experimental Demonstration of Adaptive Combinational QoT Failure Restoration in Flexible Bandwidth Networks," in National Fiber Optic Engineers Conference, Optical Society of America, 2012 OFC/NFOEC, 2012.

[21] Sambo, N., et al., "Transponder Architecture Supporting Multiwavelength Source," in 2014 International Conference on Optical Network Design and Modeling, ONDM 2014, May 2014.

[22] Barbieri, A., D. Fertonani, and G. Colavolpe, "Time-Frequency Packing for Linear Modulations: Spectral Efficiency and Practical Detection Schemes," IEEE Trans. Commun., 2009.

[23] Sambo, N., et al., "Software Defined Coderate-Adaptive Terabit/s Based on Time-Frequency Packing," in 2013 Optical Fiber Communication Conference and Exposition and the National Fiber Optic Engineers Conference (OFC/NFOEC), 2013.

[24] Xie, C., B. Zhu, and E. Burrows, "Transmission Performance of 256-Gb/s PDM-16QAM with Different Amplification Schemes and Channel Spacings," J. Lightw. Technol., Vol. 32, No. 13, July 2014, pp. 2324–2331.

[25] Cigliutti, R., et al., "16 × 125 Gb/s Quasi-Nyquist DAC-Generated PM-16QAM Transmission over 3590 km of PCSF," IEEE Photon. Technol. Lett., Vol. 24, No. 23, December 2012, pp. 2143–2146.

[26] Bosco,G., et al., "On the Performance of Nyquist-WDM Terabit Superchannels Based on PM-BPSK, PM-QPSK, PM-8QAM or PM-16QAM Subcarriers," J. Lightw. Technol., Vol. 29, No. 1, January 2011, pp. 53–61.

[27] Nespola, A., et al., "1306-km 20 x 124.8-Gb/s PM-64QAM Transmission over PSCF with net SEDP 11,300 (km)/s/Hz Using 1.15 samp/symb DAC," in Proc. 39th Eur. Conf. Exhib. Opt. Commun., September 2013, pp. 1–3.

[28] Rafique, D., et al., "Digital Preemphasis in Optical Communication Systems: On the DAC Requirements for Terabit Transmission Applications," J. Lightw. Technol., Vol. 32, No. 19, October 2014, pp. 3247– 3256.

[29] Colavolpe, G., and T. Foggi, "Time-Frequency Packing for High-Capacity Coherent Optical Links," IEEE Trans. Commun., Vol. 62, No. 8, August 2014, pp. 2986–2995.

[30] Poti, L., et al., "Casting 1 Tb/s DP-QPSK Communication into 200 GHz Bandwidth," in Proc. Eur. Conf. Opt. Commun., September 2012, pp. 1–3.

[31] Sambo, N., et al., "Programmable Transponder, Code and Differentiated Filter Configuration in Elastic Optical Networks," J. Lightw. Technol., Vol. 32, No. 11, June 2014, pp. 2079–2086.

[32] Potì, L., et al., "Sub-Nyquist Field Trial Using Time-Frequency Packed DP-QPSK Super-Channel within Fixed ITU-T Grid," Opt. Exp., Vol. 23, No. 12, June 2015, pp. 16196–16208.

[33] Meloni, G., et al., "Experimental Comparison of Transmission Performance for Nyquist WDM and Time-Frequency Packing," J. Lightwave Technol. Vol. 33, 2015, pp. 5261–5268.

[34] Chen, L., et al., "Silicon Photonics for Coherent Transmission," in Optical Fiber Communication Conference, OSA Technical Digest (online), Optical Society of America, 2016, paper Th1B.1.

[35] Wolf, S., et al., "DAC-Less Amplifier-Less Generation and Transmission of QAM Signals Using Sub-Volt Silicon-Organic Hybrid Modulators," J. Lightwave Technol., Vol. 33, 2015, pp. 1425–1432.

[36] Theodoros Santis, C., et al., "High-Coherence Hybrid Si/III-V Lasers," Proc. Natl. Acad. Sci. U.S.A., February 2014, Vol. 111, No. 8, pp. 2879-2884, DOI: 10.1073/pnas.1400184111.

[37] Yu, J., and J. Zhang, "Recent Progress on High-Speed Optical Transmission," Digital Communications and Networks, April 2016, http://dx.doi.org/10.1016/j.dcan.2016.03.002.

[38] Sambo, N., et al., "Transponder Architecture Supporting Multiwavelength Source," J. Opt. Commun. Netw., Vol. 6 , 2014 , pp. 590–600.

[39] Slavik, R. et al., "Compact Optical Comb Generator Using InP Tunable Laser and Push-Pull Modulator," IEEE Photon. Tech. L., Vol. 27, 2015, p. 217.

[40] Liang, D., et al., "150 mm InP-to-Silicon Direct Wafer Bonding for Silicon Photonic Integrated Circuits," 214th ECS Meeting, Abstract 2220.

[41] Anandarajah, P. M. et al., "Flexible Optical Comb Source for Super Channel Systems," in Optical Fiber Communication Conference and Exposition and the National Fiber Optic Engineers Conference (OFC/NFOEC), 2013, March 17–21, 2013, pp. 1–3.

[42] Goh, T., and Y. Hashizume, "Completely Cyclic Flat-Top Optical Multi/Demultiplexer for Multicarrier Transceiver," in Optical Fiber Communication Conference/National Fiber Optic Engineers Conference 2013, OSA Technical Digest (online), Optical Society of America, 2013, paper OTu3C.1.

[43] Niwa, T., et al. "Compact Integrated Tuneable Filter Utilizing AWG Routing Function and Small Switches, in Optical Fiber Communication Conference/National Fiber Optic Engineers Conference 2013, OSA Technical Digest (online), Optical Society of America, 2013, paper OW1C.2.

[44] Sun, X., et al., "Tunable Silicon Comb Filters Based on Fabry-Perot Resonators Formed by Sagnac Loop Mirrors," in Optical Fiber Communication Conference/National Fiber Optic Engineers Conference 2013, OSA Technical Digest (online), Optical Society of America, 2013, paper OTh4I.6.

[45] Sinefeld, D., S. Ben-Ezra, and D. M. Marom, "Adaptive Spectral Filtering for Nyquist-WDM Generation with Extremely High Resolution Photonic Spectral Processor," in Optical Fiber Communication Conference and Exposition and the National Fiber Optic Engineers Conference 2013 (OFC/NFOEC), March 17–21, 2013, pp. 1–3.

[46] Ip, E., et al., "Coherent Detection in Optical Fiber Systems," Opt. Express, Vol. 16, 2008, pp. 753–791.

[47] "Accelerating 400GbE Adoption with QSFP-DD," QSFP-DD MSA, http://www.qsfp-dd.com/wp-content/uploads/2017/03/QSFP-DD-whitepaper-15.pdf.

CHAPTER 10

Conclusions

Modern mobile communication systems will be remembered forever as one of the main revolutions in the history of humankind. They have led to a connected world, tearing down geographical and cultural barriers, changing our daily habits, and even modifying our brain patterns. New generations feel more natural chatting on their smartphones than talking face to face. These devices have become ubiquitous and expected, and it is easy to forget the decades of scientific research and technology development and the passion of the people who made it happen. Thinking about the impressive computing power we can carry in one hand, a smartphone is probably the most complex object the human mind ever created, embracing several cutting-edge technologies such as ultradense memories, high-speed DSP units, high-resolution photosensitive arrays, and compact high-capacity batteries, all in a volume of a few cm^3. This book focused on those aspects of a mobile communication system where optical technologies are essential. The signal generation on-air was the first topic discussed in Chapter 2, together with the rules to calculate the main physical layer parameters. The upper layers of the radio protocol stack were illustrated in Chapter 3, indicating how the combination of physical layer specifications and upper layer requirements affect the design of the optical transport network. In Chapter 4, the book offered to the reader the basic tools to plan an optical network by means of simple analytical models implementable in a spreadsheet. Then, the reader was invited to better explore the optical technologies and the different network scenarios to which those concepts apply: Chapter 4 presented the main optical transmission technologies for a mobile transport network, addressing performance issues as well as equally important aspects such as cost, power consumption, and operational issues. The case of mobile transport over a fixed access network was discussed as a relevant example of bandwidth assignment and capacity dimensioning. In Chapter 6 the importance of optical switching technologies as a necessary completion of the transmission technologies was exemplified in the ambit of the evolution of the next generation of the radio backhaul network. Starting from the state of the art of optical switches based on micromechanics or LCoS pixels, the latest research and development advances were discussed. Silicon photonics technology was referred to as a potential solution for a desired network update. Chapter 7 was the first of a series of chapters moving the focus from the optical network to optical technologies for radio equipment. In Chapter 7, the issues of analog optical transmission in radio over fiber systems were presented, outlining the analysis of the nonlinear distortions, which also apply to the air interface of a mobile system.

The subcarrier multiplexing system was used as an example to provide the reader with different analysis techniques with respect to the usual example based on binary amplitude modulation. Chapter 8 presented and discussed the synergy between integrated photonics and radio technologies to design novel beamforming antennas operative in the range of millimeter waves. Although still at the research level, these new antenna systems can be defined for multifunctional systems able to manage the wireless communications with several different frequencies and protocols Chapter 9 discussed how photonic technology advances can enhance the packet-based mobile core network through the realization of high-speed line cards with efficiently scalable cost, power consumption, and footprint. New approaches were provided to explain how to achieve better network performance with the help of optical communication and photonic integrated circuits. Silicon photonic circuits allow overcoming the limits of the electrical lines in handling high bite rates and to reduce noise and enhance system stability of radio frequency circuitry in harsh operative environments.

At the end of this long path, our wish is that the reader, whether they are a graduate, a researcher, or a professional, is inspired to further explore the potential of the new combination of optical and radio technologies to shape the next generation communication network.

List of Acronyms

2G	Second Mobile Generation
3G	Third Mobile Generation
3GPP	Third Generation Partnership Project
4G	Fourth Mobile Generation
5G	Fifth Mobile Generation
5GC	5G Core Network
3R	Reshaping, Retiming, Reamplifying
AC	Alternate Current
ACK	Acknowledgement
ADC	Analog-to-Digital Converter
AlGaAs	Aluminum Gallium Arsenide
AM	Acknowledged Mode
APD	Avalanche Photo diode
ARQ	Automatic Repeat Request
AS	Access Stratum
ASE	Amplitude Spontaneous Emission
ASIC	Application Specific Integrated Circuits
AU	Antenna Unit
AWG	Arrayed Waveguide Array
BBU	Baseband Unit
BER	Bit Error Rate
BH	Backhaul
BLER	Block Error Rate
BPSK	Binary Phase Shift Keying
BS	Base Station
CAPS	Combined Amplitude Phase Shift
CD	Chromatic Dispersion
C-RAN	Centralized RAN
CBR	Constant Bit Rate
CDMA	Code Division Multiplexing Access
CFP	Centum Form-Factor Pluggable
CN	Core Network
CNR	Carrier to Noise Ratio
CO	Central Office
CoMP	Coordinated Multipoint

CP	Cyclic Prefix
CP-OFDM	Cyclic Prefix Orthogonal Frequency Division Multiplexing
CPRI	Common Public Radio Interface
CQI	Channel Quality Indicator
CRC	Cyclic Redundancy Code
CSI	Channel State Information
CSO	Composite Second Order
CU	Central Unit
CW	Continuous Wave
CWDM	Coarse Wavelength Division Multiwavelength
DAC	Digital to Analog Converter
DBA	Dynamic Bandwidth Allocation
DBPSK	Differential Binary Phase Shift Keying
DC	Direct Current
DCF	Dispersion Compensating Fiber
DCI	Downlink Control Information
DCM	Dispersion Compensating Module
DetNet	Deterministic Networking
DFT	Discrete Fourier Transform
DFTS-OFDM	Discrete Fourier Transform Orthogonal Frequency Division Multiplexing
DGD	Differential Group Delay
DL	Downlink
DML	Directly Modulated Laser
DMT	Discrete Multitone
DPSK	Differential Phase Shift Keying
DP-QPSK	Dual Polarization QPSK
DQPSK	Differential Quadrature Phase Shift Keying
DRAN	Distributed RAN
DS	Downstream
DSF	Dispersion Compensating Fiber
DSP	Digital Signal Processing
DU	Distributed Unit
DWDM	Dense Wavelength Division Multiplexing
DwPTS	Downlink Pilot Time Slot
E/O	Electro to Optical
EAM	Electroabsorption Modulator
EDFA	Erbium-Doped Fiber Amplifier
EDGE	Enhanced Data Rates for GSM Evolution
eMBB	Enhanced Mobile Broadband
EOL	End of Life
EPC	Evolved Packet Core
EVM	Error Vector Magnitude
FAPI	Functional Application Platform Interface
FDD	Frequency Division Duplexing
FDM	Frequency Division Multiplexing
FEC	Forward Error Correction
FFT	Fast Fourier Transform

FH	Fronthaul
FWHM	Full-Wave Half Maximum
FWHM	Full-Width Half Maximum
FIR	Finite Impulse Response
FMF	Few-Mode Fiber
FM-MCF	Few-Mode Multicore Fiber
FOADM	Fixed OADM
FSK	Frequency Shift Keying
FSR	Free Spectral Range
FWM	Four-Wave Mixing
GaAs	Gallium Arsenide
GP	Guard Period
GPRS	General Packet Radio Service
GSM	Global System for Mobile Communication
HARQ	Hybrid Automatic Repeat Request
ICI	Interchannel Interference
ID	Identifier
iDFT	Inverse Discrete Fourier Transform
IEEE	Institute of Electrical and Electronics Engineers
IETF	Internet Engineering Task Force
IF	Intermediate Frequency
IFFT	Inverse Fast Fourier Transform
IM_3	Third-Order Intermodulation Product
IMT-20202	International Mobile Communications 2020
InAs	Indium Arsenide
InGaAs	Indium Gallium Arsenide
InP	Indium Phosphide
IoT	Internet of Things
IP	Internet Protocol
IQ	In-Phase In-Quadrature
ISI	Intersymbol Interference
ITU-R	International Telecommunications Union—Radiocommunication Sector
LAN	Local Area Network
LBT	Listen Before Talk
LCoS	Liquid Crystal over Silicon
LDPC	Low-Density Parity Check Code
LNA	Low-Noise Amplifier
LO	Local Oscillator
LOS	Line of Sight
LTE	Long Term Evolution
HSDPA	High Speed Downlink Packet Access
MAC	Medium Access Contro
MCF	Multicore Fiber
MCS	Modulation and Coding Scheme
MEF	Metro Ethernet Forum
MEMS	Microelectromechanical System
MIMO	Multiple Input Multiple Output

MLL	Mode-Locked Laser
mMTC	Massive Machine-Type Communication
MPLS	Multiprotocol Label Switching
MSA	Multisource Agreement
MZM	Mach-Zehnder Modulator
NACK	Nonacknowledgement
NAS	Nonaccess Stratum
NFV	Network Function Virtualization
NG-PON2	Second New Generation PON
NR	New Radio
NRZ	Non-Return to Zero
OADM	Optical Add-Drop Multiplexer
OAM	Operation, Administration and Maintenance
OBF	Optical Bandpass Filter
OBSAI	Open Base Station Architecture Initiative
ODN	Optical Distribution Node
OEO	Optical to Electrical to Optical
OFDM	Orthogonal Frequency Division Multiplexing
OFDMA	Orthogonal Frequency Division Multiplexing Access
OIF	Optical Internetworking Forum
OIP_3	Third-Order Output Intercept Point
OLT	Optical Line Terminal
OMA	Optical Modulation Amplitude
ONE	Optical Network Element
ONU	Optical Network Unit
OOK	On Off Keying
OOO	All-Optical switches
OP	Outage Probability
OSNR	Optical Signal to Noise Ratio
OSSB	Optical Single Sideband
OTN	Optical Transport Network
OXC	Optical Cross Connect
P/S	Parallel to Serial
PAM	Pulse Amplitude Modulation
PAPR	Peak to Average Power Ratio
PCB	Printed Circuit Board
PDCP	Packet Data Convergence Protocol
PD	Photodiode
PDU	Protocol Data Unit
PHY	Physical Layer
PIC	Photonic Integrated Circuit
PIN	Photo Intrinsic Diode
PLC	Planar Lightwave Circuit
PLL	Phased Locked Loop
PM-16QAM	Polarization Multiplexed-16QAM
PM-QPSK	Polarization-QPSK
PMD	Polarization Mode Dispersion
PON	Passive Optical Network

POST	Postamplifier
PRB	Physical Resource Block
PRE	Preamplifier
PSK	Phase Shift Keying
QAM	Quadrature Modulation
QoS	Quality of Service
QPSK	Quadrature Phase Shift Keying
QSFP	Quad Small Form-Factor Pluggable
QSFP-DD	QSFP Double Density
RAN	Radio Access Network
RIN	Relative Intensity Noise
RB	Radio Bearer
RBS	Rayleigh Backscattering
RE	Resource Element or Radio Equipment
REC	Radio Equipment Control
RF	Radio Frequency
RLC	Radio Link Control
ROADM	Reconfigurable OADM
RoF	Radio Over Fiber
RRC	Radio Resource Control
RRU	Remote RU
RS	Reed Solomon
RTT	Round-Trip Time
RU	Radio Unit
Rx	Receiver
RZ	Return to Zero
S/P	Serial to Parallel
SBS	Spontaneous Brillouin Scattering
SC-FDMA	Single Carrier Frequency Division Multiplexing Access
SCM	Subcarrier Multiplexing
SDAP	Service Data Adaptation Protocol
SDM	Spatial Division Multiplexing
SDN	Software-Defined Network
SDU	Service Data Unit
SER	Symbol Error Rate
SERDES	Serialization and DeSerialization
SFDR	Spurious Free Dynamic Range
SFI	Slot Format Indication
SFP	Small Form-Factor Pluggable
SMF	Single Mode Fiber
SNR	Signal to Noise Ratio
SOA	Semiconductor Optical Amplifier
SOH	Silicon-Organic Hybrid
SPM	Self-Phase Modulation
SR	Scheduling Request
SRS	Spontaneous Raman Scattering
T-CONT	Transmission Container
TACS	Total Access Communication Systems

TAE	Time Alignment Error
TB	Transport Block
TBS	TB Size
TCO	Total Cost of Ownership
TDD	Time Division Duplexing
TDM	Time Division Multiplexing
TDMA	TDM Access
TE	Time Error
TEC	Thermoelectric Cooler
TFP	Time Frequency Packing
TIA	Trans Impedance Amplifier
TM	Transparent Mode
TRL	Technology Readiness Level
TSN	Time-Sensitive Networking
TSV	Through Silicon Vias
TTI	Time Transmission Interval
TWDM	Time Wavelength Division Multiplexing
Tx	Transmitter
UE	User Equipment
UL	Uplink
UMTS	Universal Mobile Telephone System
UM	Unacknowledged Mode
UpPTS	Uplink Pilot Time Slot
URLLC	Ultra-Reliable Low-Latency Communications
US	Upstream
VCO	Voltage Controlled Oscillator
VN	Virtual Network
VOA	Variable Optical Attenuator
WDM	Wavelength Division Multiplexing
WSS	Wavelength Selective Switch
XFP	10 Gbit/s Small Form-Factor Pluggable
XPM	Cross-Gain Modulation

About the Authors

Fabio Cavaliere received his Dr. Ing. degree in telecommunications engineering from University of Pisa (Italy) in 1996. He was in the optical system design authority of Marconi (UK), from 1998 to 2006, and he is currently an expert in photonic systems and technologies at Ericsson (Sweden), where is also responsible for the standardization of optical transmission systems. In more than 20 years of professional experience, his research activity encompassed several aspects of optical networking: mobile fronthaul and backhaul, fixed fiber access, WDM systems, ultra-high-speed optical transmission, and integrated photonics. He coordinated and contributed to research projects in all these areas, in collaboration with worldwide recognized universities and research institutions. One example is the Terabit/s 1,000 km field trial on the Telstra network in Australia, carried out in collaboration with Scuola Superiore Sant'Anna and CNIT (Pisa, Italy). Cavaliere has been and is a member of the organizing and technical program committees of important international workshops and conferences on 5G radio access and optical networks such as ECOC and Globecom. He is the editor of the ITU-T Recommendation G.698.4, a member of the Board of Stakeholders of Photonics21, the European technology platform representing the photonics community since 2005, and part of the Strategic Advisory Board of the Quantum Flagship, where he contributed to define goals and structure as a member of the High-Level Steering Committee appointed by the European Commission. Cavaliere coauthored 127 filed patent applications and more than 60 publications on optical networks.

Dr Antonio D'Errico (M.Sc., Ph.D.) has been a part of Ericsson (Sweden) since 2009. D'Errico received his Ph.D. degree with honors from Scuola Superiore Sant'Anna in 2008. D'Errico has more than 15 years of experience in information technology, specializing in optical communications and fiber networks. From 2002 to 2009, he served as a researcher on new technical solutions for advanced WDM optical systems at CNIT (Pisa, Italy). He has been involved in optical access network design and contributed to the definition of a future scenario in the metro-core network, enhancing the packet-optical integration in the telecommunication transport systems. He is currently serving as a senior researcher at Ericsson Research. He has broad research interests from silicon photonics and graphene photonics technology to the design and realization of advanced solutions for the optical transport and 5G networks. He coordinates and contributes to research projects in collaboration with worldwide recognized universities and research institutions, such as the research project on bidimensional materials within the Graphene Flagship (the European initiative on graphene since 2013) in collaboration with CNIT, University of Cambridge (UK), Institute of Photonics Science (Cataluña, Spain). He is an executive board member of the Graphene Flagship. He has coorganized and cochaired international conferences, workshops, and summits on the topic of information technology and 5G evolution in conjunction with major international events, such as Graphene Week 2018, OFC 2019. He coauthored more than 90 publications in high-impact-factor international journals and conference digests (H index 21). He has published important scientific contributions with the Institute of Electrical and Electronic Engineers (IEEE), the Optical Society of America (OSA), and *Nature*. D'Errico has participated in many invited talks and plenary speech in international conference, workshops, and symposia. He holds 35 patents in optical network and advanced photonic technologies.

Index

W

Z

Recent Titles in the Artech House Applied Photonics Series

Brian Culshaw and Marc Wuilpart, Series Editors

The ABCs of Fiber Optic Communication, Sudhir Warier

Advanced Optical Communication Systems and Networks, Milorad Cvijetic and Ivan B. Djordjevic

Applications of Modern RF Photonics, Preetpaul Singh Devgan

Chemical and Biochemical Sensing with Optical Fibers and Waveguides, Gilbert Boisdé and Alan Harmer

Coherent and Nonlinear Lightwave Communications, Milorad Cvijetic

Coherent Doppler Wind Lidars in a Turbulent Atmosphere, Viktor Banakh and Igor Smalikho

Coherent Lightwave Communication Systems, Shiro Ryu

Digital Optical Measurement Techniques and Applications, Pramond Rastogi, Editor

DWDM Fundamentals, Components, and Applications, Jean-Pierre Laude

Fiber Bragg Gratings: Fundamentals and Applications in Telecommunications and Sensing, Andrea Othonos and Kyriacos Kalli

The Fiber-Optic Gyroscope, Second Edition, Hervé C. Lefèvre

Fiber-Optic Sensors for Biomedical Applications, Daniele Tosi and Guido Perrone

Finite Element Modeling Methods for Photonics, B. M. Azizur Rahman and Arti Agrawal

Frequency Stabilization of Semiconductor Laser Diodes, Tetsuhiko Ikegami, Shoichi Sudo, and Yoshihisa Sakai

Handbook of Distributed Feedback Laser Diodes, Second Edition, Geert Morthier and Patrick Vankwikelberge

Helmet-Mounted Displays and Sights, Mordekhai Velger

Introduction to Infrared and Electro-Optical Systems, Second Edition, Ronald G. Driggers, Melvin H. Friedman, and Jonathan Nichols

Introduction to Lightwave Communication Systems, Rajappa Papannareddy

Introduction to Radiometry and Photometry, Second Edition, William Ross McCluney

Introduction to Semiconductor Integrated Optics, Hans P. Zappe

Liquid Crystal Devices: Physics and Applications, Vladimir G. Chigrinov

Lithium Niobate Photonics, James E. Toney

Military Laser Technology and Systems, David H. Titterton

New Photonics Technologies for the Information Age: The Dream of Ubiquitous Services, Shoichi Sudo and Katsunari Okamoto, editors

Optical Document Security, Third Edition, Rudolf L. van Renesse

Optical FDM Network Technologies, Kiyoshi Nosu

Optical Fiber Amplifiers: Materials, Devices, and Applications, Shoichi Sudo, editor

Optical Fiber Communication Systems, Leonid Kazovsky, Sergio Benedetto and Alan Willner

Optical Fiber Sensors, Volume Three: Components and Subsystems, John Dakin and Brian Culshaw, editors

Optical Fiber Sensors, Volume Four: Applications, Analysis, and Future Trends, John Dakin and Brian Culshaw, editors

Optical Measurement Techniques and Applications, Pramod Rastogi

Optical Transmission Systems Engineering, Milorad Cvijetic

Optoelectronics for Low-Intensity Conflicts and Homeland Security, Anil K. Maini

Optoelectronic Techniques for Microwave and Millimeter-Wave Engineering, William M. Robertson

Photonic Applications for Radio Systems and Networks, Fabio Cavaliere and Antonio D'Errico

Reliability and Degradation of III-V Optical Devices, Osamu Ueda

Signal Processing and Performance Analysis for Imaging Systems, S. Susan Young, Ronald G. Driggers, and Eddie L. Jacobs

Smart Structures and Materials, Brian Culshaw

Substrate Surface Preparation Handbook, Max Robertson

Surveillance and Reconnaissance Imaging Systems: Modeling and Performance Prediction, Jon C. Leachtenauer and Ronald G. Driggers

Wavelength Division Multiple Access Optical Networks, Andrea Borella, Giovanni Cancellieri, and Franco Chiaraluce

For further information on these and other Artech House titles, including previously considered out-of-print books now available through our In-Print-Forever® (IPF®) program, contact:

Artech House
685 Canton Street
Norwood, MA 02062
Phone: 781-769-9750
Fax: 781-769-6334
e-mail: artech@artechhouse.com

Artech House
16 Sussex Street
London SW1V 4RW UK
Phone: +44 (0)20 7596-8750
Fax: +44 (0)20 7630-0166
e-mail: artech-uk@artechhouse.com

Find us on the World Wide Web at: www.artechhouse.com